# A Case Study Comparing Two Types Of System Development Projects And Their Implementation

## By

## Dr. Jennifer W. Gilmore, Ph.D.

Kennedy-Western University

ISBN: 1-4107-0684-2 (e-book)
ISBN: 1-4107-0685-0 (Paperback)

Library of Congress Control Number: 2002096820

This book is printed on acid free paper.

Printed in the United States of America
Bloomington, IN

1stBooks – rev. 01/04/03

This book describes a case study of two systems development projects completed in the 1990s at New York City's Human Resources Administration by Management Information Systems staff. This researcher was the project leader and senior systems analyst on both projects. The two methods of systems development, the systems development life cycle and prototyping, are described in relation to how these two methods are used in the development of the two computer systems. Other methods of systems development are described in the literature review. The high points of these two projects are mentioned and the reasons for the success of these two systems development projects are noted. Also important was the effort put into the development by the developers and the cooperation of users and management in order to produce two computer systems that are still up and running successfully today and will in the future.

# TABLE OF CONTENTS

# INTRODUCTION

## Chapter 1

### Introduction/Importance of the Study

This study deals with the development and implementation of two systems using two methods of system development. The importance of this study is that no matter what type of system development used to develop and implement a system in the end it is the procedure followed. Also important is the effort put into the development by the developers and the cooperation of management in order to produce a computer system that will run successfully for many years. Also important is the return on investment.

According to Post and Anderson (2000), the bottom line is that managers often have to decide which method of systems development is most suitable for developing a particular system. If the project is large or expensive, and affects important assets of the company, the choice of Systems Development Life Cycle (SDLC) is fairly clear-cut. If a user simply needs computations for a one-time decision, it is faster and cheaper to have the user create the model using a spreadsheet. However, in many cases, there is not an obvious answer. A project might start out small, then suddenly grow. Similarly, a spreadsheet that was created by a user for one purpose might be passed around the company and used by others to make crucial decisions.

There is not much difference between projects created by end users and those that use prototyping. In a sense, virtually any project could be started by end users. The trick is to learn when to call for help by understanding the limitations of the methods. If the project is used by many people, affects critical portions of the business, or grows beyond the capabilities of prepackaged software, then Management Information Systems (MIS) programmers can be called in to expand it, test it, create documentation, make it more flexible, or create an entirely new system.

Modern MIS departments have devised many ways to help users develop new systems other than the traditional analysts and programmers working on SDLC projects. Two methods are moving MIS employees out to the departments and the creation of an information center. In both cases, the goal is to have MIS employees responsible for assisting users. This assistance can be offered in many forms, such as helping decide which hardware and software to purchase, answering questions about software, offering classes in using software and creating systems, testing user-development systems for errors and incompatibilities, and building prototypes.

Turban and Aronson (1998) stated that the impact of computer technology on organizations and society is increasing as new technologies evolve and existing technologies expand. The interaction and cooperation between people and machines are rapidly growing to cover more and more aspects of organizational activities. From traditional uses in payroll and bookkeeping functions, computerized systems are now penetrating complex managerial areas ranging from the design and management of automated factories to the evaluation of proposed mergers and acquisitions. Nearly all business executives say that information technology is vital to the business and they are using technologies extensively.

Turban and Aronson (1998) continued to say that computer applications are moving from transaction or backroom processing and monitoring activities to problem analysis and solution applications. Topics such as data access, online analytical processing, and use of the Internet and Intranets for decision support are becoming the cornerstones of modern management. There is a trend to provide managers with information systems that can assist them directly in their most important task: making decisions.

## Purpose of the Study

The purpose of this study is to describe in detail two methods of system development, the SDLC and Prototyping, used in the development of two successfully implemented business systems and to describe the reasons for their successful development and successful implementation and use.

## Problem Statement

The systematic automation of the workplace is essential for businesses to realize efficiency, profits, and increased levels of production to achieve their goals and aspirations more quickly and at less cost. The work of employees could be transformed into business computer systems which employees could use to do their daily work at faster speeds with accuracy, reliability, less cost, and with greater production and profit realization. This study will investigate through the use of a case study two types of systems development used at New York City Human Resources Administration by the Management Information Systems personnel, one of which is this researcher as a computer specialist, senior systems analyst, and project manager. The two types of systems development being studied are the systems development life cycle and prototype development to see if the advantages of these two

3

methods of systems development far outweigh the disadvantages of no system development. This study will also look at other methods of systems development mostly garnered from the review of the literature.

The research question for this study will therefore be: Are there more advantages than disadvantages in the development of business software computer systems using the systems development life cycle method and the prototype development method or are there other methods of systems development that are better suited to business environments?

## Scope of the Study

This case study encompasses the work of a senior systems analyst (this researcher) in the MIS Department of the City of New York, Human Resources Administration. Two systems were developed for use by the employees of two departments of HRA in the mid- to late-1990s using two different development strategies.

## Rationale of the Study

This case is worthy of in-depth study because it will contribute to our knowledge about computer systems development in business and in government. It will describe the similarities and differences between systems development using the systems development life cycle versus prototype systems development. It will also reveal the strong points about the two methods of development for the two systems that were successfully developed and are still up and running successfully today.

## Definition of Terms

HRA      Human Resources Administration Agency

MIS      Management Information Systems

NYC      New York City

## Overview of the Study

The first chapter of this study will describe the introduction and importance of the study. Other topics described in Chapter 1 include the background of the problem and the purpose of this study. Chapter 2 will present the literature review of the different methods of system development and other related literature. Chapter 3 will describe the methodology of the study. Chapter 4 describes the actual case study with the presentation of the data and analysis of the data. Chapter 5 describes the summary, answer to the research question, the discussion, the recommendations, and conclusion of this study.

## Summary

This chapter presented the introduction to this study and the importance, purpose and rationale for this study among other things. Chapter 2 will present a review of the related literature for this study.

# REVIEW OF RELATED LITERATURE

## Chapter 2

### Introduction

The related literature encompasses all methods of system development found in the literature. These include the System Development Life Cycle, Prototyping, Rapid Application Development, Object-Oriented Development, End-User Development, Expert System Development, Neural Network Application Development, and the development of Executive Information Systems. These different approaches as described by different authors are included in this chapter. Reasons for success or failure of systems are also covered.

### Methods of Creating Information Systems

Post and Anderson (2000) indicate that there are three basic techniques to create information systems to support business needs. They are systems development life cycle, prototyping, and end-user development. The systems development life cycle (SDLC) is the most formal approach and is used by large organizations that develop several systems to coordinate team members, evaluate progress, and ensure quality development. Prototyping involves creation of a working version of the system as quickly as possible that may not contain all the components. End-user development relies on users who have some computer skills to create their own systems.

Turban and Aronson (1998) similarly compared the Life Cycle Development method versus Prototyping in the development of decision support systems. Turban and Aronson stated that Decision Support Systems (DSS) construction could be performed in several different ways. Differentiation is made between the two extremes: the life-cycle approach and the evolutionary prototyping approach (iterative process). DSS designers have recognized a need to depart from the traditional system development life cycle (SDLC) process. Instead a prototyping approach called the evolutionary prototyping process, iterative process, or just prototyping has been adopted.

Turban and Aronson (1998) continued to explain that the prototyping approach aims at building a DSS in a series of short steps with immediate feedback from users to ensure that development is proceeding correctly. Therefore, DSS tools must permit changes to be made quickly and easily. The iterative process includes the following four tasks: (1) Select an important subproblem to be built first; (2) Develop a small but usable system for the decision-maker; (3) Evaluate the system constantly; and (4) Refine, expand, and modify the system in cycles.

McLeod (1995) put the System Life Cycle, prototyping, and Rapid Application Development methods in perspective by stating that the SLC, prototyping, and Rapid Application Development (RAD) are all methodologies. They are recommended ways of implementing a computer-based system. The SLC is an application of the systems approach to the problem of computer system implementation, and contains all of the elements, beginning with problem identification and ending with system use.

Prototyping is an abbreviated form of the systems approach that focuses on the definition and satisfaction of user needs. Prototyping can exist within the SLC. In fact, many prototyping efforts may be required during the process of developing a single system.

RAD is an alternative approach to the design and implementation phases of the SLC. The main contribution of RAD is the speed with which systems are put into use, achieved primarily through the use of computer-based tools and specialized project teams.

Of all the methodologies, the SLC has been around the longest and will most likely continue to provide the basis for much development work. Prototyping is also well established, and will continue to be used for those projects where definition of user needs is difficult. RAD is just now coming onto the scene, and its future is less certain.

## The Systems Development Life Cycle Approach

The SDLC approach encompasses five basic stages (1) feasibility and planning, (2) systems analysis, (3) systems design, (4) implementation, and (5) maintenance and review. A description of this approach as described by Post and Anderson (2000) follows below.

### Feasibility and Planning

The first step in systems analysis is the feasibility study. A feasibility study is a quick examination of the problems, goals, and expected costs of the system. The objective is to determine whether the problem can reasonably be solved with a computer system. The scope of the project is determined to gain a better idea of the costs, benefits, and objectives.

The feasibility study is typically written so that non-programmers can easily understand it. It is used to sell the project to upper management and as a starting point for the next step. It is also used as a reference to keep the project on track, and to evaluate the progress of the

MIS team. Projects are typically evaluated in three areas of feasibility: economics, operations, and technical. When the proposal is determined to be feasible, the MIS team leaders are appointed and a plan and schedule are created. The schedule contains a detailed listing of what parts of the project will be completed at each time. It is of course extremely difficult to estimate the true costs and completion dates. However, the schedule is an important tool to evaluate the status of the project and the progress of the MIS teams.

**Systems Analysis**

Once a project has been shown to be feasible and it is approved, work can begin on a full-fledged analysis. The first step is to determine how the existing system works and where the problems are located. The technique is to break the system into pieces. Smaller pieces are easier to understand and to explain to others. Also, each piece can be assigned to a different MIS team. As long as they work from the same initial description and follow all of the standards, the resulting pieces should fit back together. Of course, it still takes time and effort to integrate all of the pieces.

Diagrams are often created to illustrate the system. The diagrams are used to communicate among analysts and users, other analysts, and eventually the programmers. Data flow diagrams are a common method to display the relationships that were determined during systems analysis. The diagrams represent a way to divide the system into smaller pieces.

Graphics tools provide a useful way to communicate with the user and to document the user requirements. Producing, changing, and storing documentation can be a significant problem. They do not speed up the development process. Yet these tools are necessary because they make it easier for the user to control the final result. One increasingly common solution is to keep all of the documentation on the computer. This method reduces the costs, makes it easier for everyone to share the documentation, and ensures that all users have up-to-date information for the system.

At the end of the analysis phase, the MIS team will have a complete description of the business requirements. The problems and needs are documented with text, data flow diagrams, and other figures depending on the methodology followed.

**Systems Design**

The third major step of the SDLC is to design the new system. During this step, the new system is typically designed on paper. The objective of systems design is to describe the new system as a collection of modules or subsystems. By subdividing the total project, each portion can be given to a single programmer to develop. As the pieces are completed, the overall design ensures that they will work together. Typically, the diagrams created during the analysis phase can be modified to indicate how the new system will work. The design will list all of the details, including data inputs, system outputs, processing steps, database designs, manual procedures, and feedback and control mechanisms. Backup and recovery plans along with security controls will be spelled out to ensure that the database is protected.

In traditional SDLC methods, managers and users will be shown various components of the system as they are completed. The managers will have to sign off on these sections to indicate that they meet the user needs. This signature is designed to ensure that users provide input to the system. If there are many diverse users, there can be major disagreements about how the system should function. Signoffs require users to negotiate and formally agree to the design. It is relatively easy to make design changes at this stage. If everyone attempts to make changes at later stages, the cost increases dramatically.

In terms of physical design, some of the hardware and software will be purchased. Programmers will write and test the program code. In most large projects, the actual coding takes only 15 to 30 percent of the total development time. Initial data will be collected or transferred from existing systems. Manuals and procedures will be written to instruct uses and system operators on how to use the system.

Once the designer has created base modules and sample inputs and outputs, the users are invited to a structured walkthrough. A structured walkthrough is a review process where the objective is to reveal problems, inaccuracies, ambiguities, and omissions in the systems design before the program code is finalized. The users are presented with a prototype or mockup of the proposed system. It is easier to spot problems and make suggestions by observing how the actual system might appear.

Design tools can be used to create prototypes of major system elements. For example, a designer can quickly piece together displays that illustrate how each screen might look and how the user will see the system. The number of walkthroughs used depends on the amount of time users and programmers can spend on reviewing the designs. The walkthroughs also provide management with feedback regarding the time schedule and anticipated costs of the project, because they are often scheduled in the original feasibility study.

The output of the design stage consists of a complete technical specification of the new system. It includes as many details as possible—sometimes leading to thousands of pages (or computer files) of description.

One of the difficulties in the design stage is sometimes called 'creeping elegance'. As the system is being built analysts, programmers, and users all want to include additional features. Although many of the features are good ideas, the continual evolution of the system causes additional delays. It also complicates testing, because changes in one section can affect the rest of the system.

## Systems Implementation

Systems implementation involves installation and changeover from the previous system to the new one, including training users and making adjustments to the system. Many nasty problems can arise at this stage. You have to be extremely careful in implementing new systems. First, users are probably nervous about the change already. If something goes wrong, they may never trust the new system. Second, if major errors occur, you could lose important business data.

A crucial stage in implementation is final testing. Testing and quality control must be performed at every stage of development, but a final systems test is needed before staff entrust the company's data to the new system. Occasionally, small problems will be noted, but their resolution will be left for later. In any large system, there are bound to be errors and changes. The key is to identify them and determine which ones must be fixed immediately. Smaller problems are often left to the software maintenance staff.

Change is an important part of MIS. Designing and implementing new systems often causes changes in the business operations. Yet, many people do not like changes. Changes require learning new methods, forging new relationships with people and managers, or perhaps even loss of jobs. Changes exist on many levels: in society, in business, and in information systems. Changes can occur because of shifts in the environment, or internal change agents can introduce them. Left to themselves, most organizations will resist even small changes. Change agents are objects or people who cause or facilitate changes. Sometimes it might be a new employee who brings fresh ideas; other times changes can be mandated by top-level management. Sometimes an outside event such as arrival of a new competitor or a natural disaster forces an organization to change. Whatever the cause, people tend to resist change. However, if organizations do not change, they cannot survive. The goal is to implement systems in a manner that recognizes resistance to change but encourages people to accept the new system. Effective implementation involves finding ways to reduce this resistance. Sometimes, implementation involves the cooperation of outsiders such as suppliers.

An important process in reducing resistance to change is to involve the users in the design of the new system. A key component of any implementation strategy is user education and training. It is important that systems be flexible so the users can adjust them. Simple things like being able to set screen colors, mouse sensitivity, and keyboard speed can be critical to keeping users happy. They are also useful for users with physical challenges.

It is also important to recognize how the system will affect the business. Computer systems often change the way the business operates. Also, encourage users to change. A common management technique is to make sure that the payment system incentives match the goals of the organization. Even if the business operations are not substantially altered, implementing a new system can cause reduced productivity while employees learn to operate the new system. A common solution is to provide additional training to help the users learn the system faster.

Because implementation is so important, several techniques have been developed to help implement new systems. Direct cutover is an obvious technique, where the old system is simply dropped and the new one started. If at all possible, it is best to avoid this technique, because it is the most dangerous to data. If anything goes wrong with the new system, you run the risk of losing valuable information because the old system is not available. In many ways, the safest choice is to use parallel implementation. In this case, the new system is introduced alongside the old one. Both systems are operated at the same time until you determine that the new system is acceptable. The main drawback to this method is that it can be expensive because data has to be entered twice. Additionally, if users are nervous about the new system, they might avoid the change and stick with the old method. In this case, the new system may never get a fair trial. Several intermediate possibilities are called phased implementation.

**Maintenance**

Once the system is installed, the MIS job has just begun. Computer systems are constantly changing. Hardware upgrades occur continuously, and commercial software tools may change every year. Users change jobs. Errors may exist in the system. The business changes, and management and users demand new information and expansions. All of these actions mean the system needs to be modified. The job of overseeing and making these modifications is called software maintenance.

The pressures for change are so great that in most organizations today as much as 80 percent of the MIS staff is devoted to modifying existing programs. These changes can be time consuming and difficult. Teams of programmers and analysts created most major systems over a long period. In order to make a change to a program, the programmer has to understand how the current program works. Because many different people with varying styles wrote the program, it can be hard to understand. Finally, when a programmer makes a minor change in one location it can affect another area of the program, which can cause additional errors or necessitate more changes.

**Evaluation**

Evaluation is a difficult issue. How can a manager tell the difference between a good system and a poor one? In some way, the system should decrease costs, increase revenue, or provide a competitive advantage. Although these effects are important, they are often subtle and difficult to measure. The system should also be easy to use and flexible enough to adapt to changes in the business. If employees or customers continue to complain about a system, it should be reexamined.

A system also needs to be reliable. It should be available when needed and should produce accurate output. Error detection can be provided in the system to recognize and avoid common problems. Similarly, some systems can be built to tolerate errors, so that when errors arise, the system recognizes the problem and works around it. For example, some computers exist today that automatically switch to backup components when one section fails, thereby exhibiting fault tolerance.

An important concept for managers to remember when dealing with new systems is that the evaluation mechanism should be determined at the start of the project. Far too often, the question of evaluation is ignored until someone questions the value of the finished product.

Larman (1997) in a similar way to Post and Anderson (2000) described the phases of a development cycle as follows:

The Analyze phase of a development cycle emphasizes an investigation of the problem and requirements analysis. The Design phase of a development cycle emphasizes defining a logical solution. The Construct phase involves implementation of the design in software and hardware. Although the Test phase is shown as a final step within a development cycle, it is also recommended as an ongoing activity during the Construct phase.

McLeod (1995) similarly to Larman (1997) and Post and Anderson (2000) described the system life cycle (SLC) as the evolutionary process that is followed in implementing a computer-based information system or subsystem. The SLC consists of a series of tasks that closely follow the steps of the systems approach. Since the tasks follow an orderly pattern and are performed in a top-down fashion, the SLC is often referred to as the waterfall approach to systems development and use.

McLeod (1995) explained that the first four life cycle phases are planning, analysis, design, and implementation. These phases are jointly called the system development life cycle (SDLC). The fifth phase is the use phase, which lasts until it is time to redesign the system. Redesign requires that the cycle be repeated.

McLeod further explained that many firms establish a special committee, below the level of the executive committee that assumes responsibility for overseeing all of the systems projects. When the purpose of a committee is to provide ongoing guidance, direction, and control, it is called a steering committee. When a firm establishes a steering committee for the purpose of directing the use of the firm's computing resources, the name MIS steering committee is used. Permanent members of the MIS steering committee invariably include top executives. Temporary members include lower-level managers and consultants who participate during the time that their expertise is needed.

The MIS steering committee seldom gets directly involved with the details of the work. That responsibility goes to project teams. A project team includes all of the persons who participate in the development of a computer-based system. A team might have as many as a dozen members, consisting of some combination of users, information specialists, and perhaps an internal auditor. The auditor ensures that the system design satisfies certain requirements in terms of accuracy, controls, security and auditability. A project leader who provides direction throughout the life of the project directs the team activity. Unlike the MIS steering committee, the project team is not ongoing in that it is usually disbanded when implementation is completed.

McLeod (1995) in a similar fashion to Larman (1997) and Post and Anderson (2000) described the steps of the planning phase as follows. Recognize the problem; Define the problem; Set system objectives; Identify system constraints; Conduct a feasibility study; Prepare a system study proposal; Approve or Disapprove the study project; and Establish a control mechanism.

The analysis phase includes the following steps: Announce the system study; Organize the project team; Define information needs; Define system performance criteria; Prepare the design proposal; and Approve or Disapprove the design project. The design phase includes the following: Prepare the detailed system design; Identify alternative system configurations; Evaluate alternative system configurations; Select the best configuration; Prepare the implementation proposal; and Approve or disapprove the system implementation.

The implementation phase includes the following: Plan the implementation; Announce the implementation; Obtain the hardware resources; Obtain the software resources; Prepare the database; Prepare the physical facilities; Educate the participants and users; and Cutover to the new system. The use phase includes the following steps: Use the system; Audit the system; and Maintain the system.

## Strengths and Weaknesses of SDLC

According to Post and Anderson (2000), the primary purpose of the SDLC method of designing systems is to provide guidance and control over the development process. SDLC management control is vital for large projects to ensure that the individual teams work together. There are also financial controls to keep track of the project expenses. The SDLC steps are often spelled out in great detail. The formality makes it easier to train employees and to evaluate the progress of the development. It also ensures that steps are not skipped—such as user approval, documentation, and testing. For large, complex projects, this degree of control is necessary to ensure the project can be completed. Another advantage of the SDLC is that by adhering to standards while building the system, programmers will find the system easier to modify and maintain later. The internal consistency and documentation make it easier to modify. With 80 percent of MIS resources spent on maintenance, this advantage can be critical.

In some cases the formality of SDLC also causes problems with projects that are hard to define. SDLC works best if the entire system can be accurately specified in the beginning. That is, users and managers need to know exactly what the system should do long before the system is created. That is not a serious problem with transaction-processing systems.

Although some large projects could never have been completed without SDLC, its rigidity tends to make it difficult to develop many modern applications. Additionally, experience has shown that it has not really solved the problems of projects being over budget and late. As a result of this criticism, many people are searching for alternatives. One possibility is to keep the basic SDLC in place and use technology to make it more efficient. Other suggestions have been to replace the entire process with a more efficient development process, such as prototyping. Consider the assistance of technology first.

Larman (1997) described the disadvantages of a waterfall life cycle like SDLC as follows:

The singular characteristic of a waterfall life cycle of development is that it involves a single pass of analysis, design, and construction; one does all the analysis, then all the design, then the coding, and finally, all the testing. Some of its flaws include (1) front-loading of tackling complexity—complexity overload (2) delayed feedback and (3) specifications frozen early, while business environment changes.

## Prototyping Approach

Post and Anderson (2000) described prototyping as follows:

Prototyping has been proposed as a method to use for systems that are not overly complex and do not involve too many users or analysts. Just as automobile engineers design prototypes before attempting to build the final car, MIS programmers can build early versions of systems. These systems are continually modified until the user is satisfied.

The first step in designing a system via prototyping is to talk with the user. The analyst then uses a fourth-generation language and a DBMS to create approximately what the user wants. This first step generally requires only a couple of weeks. The business user then works with the prototype and suggests changes. The analyst makes the changes and this cycle repeats until the user is satisfied or decides that the system is not worth pursuing. The emphasis is on getting a working version of the system to the user as fast as possible. And, Even if it does not have all the details.

The cycle involved in prototyping is as follows. Prototyping typically involves one user and one developer. The developer interviews the user and designs an initial system using a DBMS. The user works with the prototype and suggests changes. This process repeats until the user or developer is satisfied or gives up.

The major advantage of prototyping is that users receive a working system much sooner that they would with the SDLC method. Additionally, the users have more input so they are more likely to get what they wanted. A system designed with the prototyping method is much easier to change because it was designed to be modified from the start.

In the early stages of design, a prototype is analogous to the scale models that automobile and aircraft designers use for wind-tunnel tests or to the cardboard scale models that architects use to show potential buyers what the finished building will look like. The initial mockup may not actually have any computer code. It might be a collection of input screens and reports that look like they were produced by the computer system. The primary value of these simulated results is to enable the user to actually see the planned results. Users often have difficulty telling designers what they want beforehand. They can react to sample output by saying what they like or dislike, or what needs to be added to the output to make it more useful to the end user.

With today's latest development tools, however, the prototype may be more than a mockup. It may be used to create a substantial portion of the final code. Using Computer Aided Systems Engineering (CASE) tools and database management systems, error-free code can be generated almost instantly from the prototype descriptions. These tools also build documentation automatically, making it easier for the designer and the maintenance programmer.

Tools available for prototyping include presentation tools, which can show a series of computer screens and demonstrate the look and feel of the finished product. More powerful fourth-generation languages and database management systems can generate actual applications including input files, interactive screens, reports, and executable code without requiring programming. Other specialized tools such as expert system shells enable the designer to work with complex logical rules to create a finished product for certain types of applications.

The process of building a preliminary prototype, trying it out, refining it, and trying again is referred to as the interactive process of systems development. In the traditional SDLC approach, there was some iteration, but it is limited to a defined stage. The formal sign-off procedure was instituted specifically to restrict changes beyond some point in the development cycle. With the prototyping approach, iteration is planned and modern prototyping tools make it easy to modify the application. Further, because the prototyping tools allow a preliminary model to be created much faster, the user gets a working model much earlier in the design process. Therefore, the repeated changes are less likely to delay completion.

Prototyping is especially helpful in situations where there will be heavy user interaction with the system, where the needed output is uncertain, and in some decision support applications where the logic is hard to determine in advance. Prototyping fosters a spirit of experimentation. It does not matter if there is some uncertainty regarding the path to the desired results or even uncertain about the desired results. Prototyping enables you to try out a possible solution to see how it flies and to keep repeating this process until users are satisfied.

McLeod (1995) in a similar fashion to Post and Anderson (2000) described prototyping. He similarly stated that a prototype provides developers and potential users with an idea of how the system in its completed form will function. McLeod, however, distinguished between two types of prototypes. A Type I prototype eventually becomes the operational system. A Type II prototype is a throwaway model that serves as the blueprint for the operational system.

The development of a Type I prototype involves the following steps: Identify user needs; Develop a prototype; Determine if the prototype is acceptable; and Use the prototype. The development of a Type II prototype includes the following steps: The first three are the same as for a Type I prototype, the next steps are as follows: Code the operational system; Test the operational system; Determine if the operational system is acceptable; and Use the operational system.

## Advantages of Prototyping

Turban and Aronson (1998) stated that the following are some of the major advantages of prototyping: Short development time; Short user reaction time (feedback from user); Improved user understanding of the system, its information needs, and its capabilities; and Low cost.

McLeod (1995) similarly indicated that both users and information specialists like prototyping for the following reasons: Communications between the systems analyst and user are improved; The analyst can do a better job of determining the user's needs. The user plays a more active role in system development. The information specialists and the user spend less time and effort in developing the system; and Implementation is much easier because the user knows what to expect. These advantages enable prototyping to cut developmental costs and increase user satisfaction with the delivered system.

## Drawbacks and Complications of Prototyping

Post and Anderson (2000) indicated that prototyping does have some drawbacks and complications. First, problems arise if several users are involved. Imagine what happens if each user wants different changes and they cannot agree on how the system should behave. One of the reasons for user sign-offs in SDLC is to force users to agree before you create the system. For large projects, several different prototypes of reports and input screens could be used to help users evaluate the various choices.

There are some more subtle drawbacks to prototyping. Picture a MIS manager where the analysts all use prototyping exclusively. Most of the time the analysts are talking with the users and making simple changes. How would it be known when the project is finished? How would it be known whether the analyst is a hard worker or just making endless minor changes? How would the user be told to stop making trivial changes? Under the SDLC method, there are predefined targets (milestones) that it is known exactly what each person is supposed to accomplish. Which method would a manager responsible for completing projects and allocating raises prefer?

Turban and Aronson (1998) described disadvantages and limitations of prototyping. When such an approach is used, the gains obtained from cautiously stepping through each of the system's life-cycle stages might be lost. These gains include a thorough understanding of the information system's benefits and costs, a detailed description of the business's information needs, an easy to maintain information system design, a well-tested information system, and well-prepared users.

McLeod (1995) similarly described the following potential pitfalls of prototyping: The haste to deliver the prototype may produce shortcuts in problem definition, alternative evaluation, and documentation. The term "quick and dirty" has been used to describe some prototyping efforts. The users may get so excited about the prototype that they have unrealistic expectations of the operational system. Type I prototypes might not be as efficient as systems coded in a programming language. The computer-human interface provided by certain prototyping tools may not reflect good design techniques.

## Rapid Application Development (RAD) Approach

Post and Anderson (2000) described Rapid Application Development as follows:

Rapid application development (RAD) bears some similarities to prototyping, but it attempts to maintain more control. The objective is to follow a more formal project approach but find the steps that can be reduced or performed at the same time. Using high-level languages, database systems, project management software, workgroup software, and CASE tools, highly trained programmers can build systems in a matter of weeks or months.

McLeod (1995) similarly described the rapid application development methodology as a methodology that has the same objective of speedy response to user needs as does prototyping but is broader in scope. RAD is an integrated set of strategies, methodologies, and tools that exists within an overall framework called information engineering. Information engineering begins at the executive level, with strategic information resources planning applied to the entire enterprise. Next, each business unit within the firm is subjected to business area analysis (BAA) to define the activities or processes and data that are necessary for the unit to function as intended. With the BAA completed, rapid application development can proceed.

RAD requires four essential ingredients: management, people, methodologies, and tools.

**Management**.    Management, especially top management, should be experimenters, who like to do things a new way, or early adapters, who quickly learn how to use new methodologies. Management should be fully supportive of RAD and provide a work environment that makes the activity as enjoyable as possible.

**People**.    Rather than utilize a single team to perform all of the SLC activities. RAD recognizes the efficiencies that can be achieved through the use of several specialized teams. There can be teams for requirements planning, user design, construction, user review, and cutover. Members of these teams use masters of the methodologies and tools that are required to perform their specialized tasks.

**Methodologies**.    The basic RAD methodology is the RAD life cycle, which consists of four phases: (1) requirements planning, (2) user design, (3) construction, and (4) cutover. These phases, like the SDLC, reflect the systems approach. Users play key roles in each phase, participating with information specialists.

**Tools**.    RAD tools consist mainly of fourth-generation languages and CASE tools that facilitate prototyping and code generation. Fourth-generation languages enable information specialists or users to generate computer code without using a conventional programming language.

## Object-Oriented Development Approach

Post and Anderson (2000) described the object-oriented development approach as follows:

The concept of object-oriented development has received considerable attention during the past few years. In some ways, the base design techniques are not much different from traditional SDLC techniques. In other ways, object orientation requires a completely new way of thinking about systems development. The ultimate goal of the object-oriented approach is to build a set of reusable objects and procedures. The idea is that eventually, it should be possible to create new systems or modify old ones simply by plugging in a new module or modifying an existing object.

An object can be anything from an icon on a computer screen to an accounting statement. Objects have a set of characteristics or attributes, and methods or operations that can be performed on objects. Properties and methods are inherited. New objects based on other objects acquire the same properties and methods. Designers need only define the additional properties and functions that make the new objects different. Once the base classes (collections of objects) are defined, new objects can be created with minimal effort.

One key difference between object orientation and other development methods is the way processes or functions are handled. With objects, all functions are embedded in the definition of the object—the object comes first. The object approach reverses the treatment of processes and data. With SDLC, the emphasis is on processes, and data (attributes) is passed between processes.

One goal of an object-oriented approach is to create a set of information system building blocks. These objects and procedures could be purchased from commercial software companies (such as a spreadsheet from Microsoft or a database system from Oracle). MIS programmers or consultants can create additional objects tailored for your specific company or department. Once the basic blocks are in place, end users or MIS analysts can select the individual pieces to create a complete system. Hence, less time is needed for implementation as long as the analysis and design are performed carefully. On the other hand, the up-front costs of designing and building these objects can be quite high. Additionally, the tools and techniques tend to require substantial retraining of the existing MIS staff. Both of these types of costs have caused some companies to avoid object-oriented methods.

Although object-oriented techniques are still being developed, many companies are already headed in this direction. One effect of this type of development is that much of the traditional programming is moved to commercial companies that specialize in creating software components. Other companies hire analysts who focus on the business needs and put together the appropriate modules. These analysts need to understand business, be able to identify problems, and know how information systems can be used to solve them.

Post and Anderson (2000) further explained the object-oriented design approach in the following:

One way to begin your analysis of a business is to focus on the business objects: what they are and what they do. Objects could be anything from people to raw materials to data files or schedules. The key to object-oriented design is to focus on defining what an object is and what it can do. A class is a generic description of a set of objects.

Objects are defined by a set of properties (or attributes). The properties define the object. They also represent data that will be collected for each object. Each object also has functions which describe actions that can be performed by the objects and define how to alter the object. With the object-oriented approach the properties and functions are combined into the definition of the object. The goal is to describe a system so that if you change a function, you only have to change one object. All of the other objects and processes remain the same.

Objects are related to each other. Typically there is a base class of objects and other objects are derived from the base definitions by adding properties and altering functions. This process results in an object hierarchy.

Another aspect of modeling objects is that they are often used in an event-driven approach. When some business event occurs, an object function is called or a property is modified. Possible events need to be thought up and how they influence the objects being controlled.

To solve business problems, how the business operates first has to be understood. The basic idea is that systems consist of smaller, interdependent subsystems. Each subsystem can be broken into smaller sections with more details. By examining each piece and its interactions, it is easier to determine the cause of problems and to derive a solution.

Several techniques have been created to help analyze systems. A useful new method is based on the unified modeling language (UML). UML was designed to assist in creating object-oriented information systems. UML asks that the primary objects in the system be identified in terms of their properties and the methods or functions that they can perform.

Most of today's information systems rely heavily on database management systems to collect and share the underlying data. An important strength of this approach is that you rarely need to worry about how data gets from one point in the company to another. Managers simply retrieve the data from the database as needed. The UML design techniques are useful in this environment because they focus on collecting and manipulating the data stored in each object.

Gossain (1998) agrees with Post and Anderson (2000) by stating that moving to object orientation requires software development communities to open up their analysis-and-design process by involving nontechnical people in the process. Analysts need to involve people from the business domain throughout the process, and in some cases this can be a major sticking point.

In a similar vein to Post and Anderson (2000), Gossain (1998) stated that architecture is key. Objects facilitate more flexible architectures. Before designing the system the architecture must be defined and then continually revisited throughout the process. Objects allow us to describe, discuss, and reason about software architecture in a way that previous paradigms never did. This is not to say that architecture was not an issue until objects came around. The spread of objects has been in step with the increasing awareness of architecture. It is an important aspect of system construction. It is the unique packaging of data and behavior into objects that allow us to discuss the architectural issues surrounding system construction. Thinking of subsystems and systems as objects allows us to apply object ideas to large grain concepts.

Gossain (1998) explained that although it is true that objects are the building blocks of the system, merely following an object-oriented analysis-and-design method does not guarantee a 'good' architecture. Each project must have an architect, one who is responsible for ensuring the macro-structure of the system. It is important that the high-level architecture has been defined by the time design comes around. After that, as the design proceeds, the architecture must be continually revisited. It will, of course, change slightly, but the design should adhere to the principles of the architecture.

Gossain (1998) continued to explain that when making the transition, management education, and then buy-in, is essential. Managers at all levels need to be educated in order to appreciate fully the impact of objects. The need for leadership is another aspect worth mentioning. It is important to distinguish between management and leadership. Using object technology, and learning from the projects taken up can require strong leadership. When selecting the first pilot project, choose one that has high visibility in the business community and one that will make a difference.

## Advantages of the Object Oriented Approach

Larman (1997) stated that in contrast to the waterfall life cycle like SDLC, an iterative life cycle like object oriented development is based on successive refinement of a system through multiple cycles of analysis, design and construction. Each cycle tackles a relatively small set of requirements and the system grows incrementally as each cycle is completed. Hence this process is characterized as iterative and incremental; it includes the following benefits:

1. The complexity is never overwhelming, because only manageable units of complexity are tackled in a cycle.

2. Early feedback is generated, because implementation occurs rapidly for a small subset of the system. This feedback can inform and enhance the analysis of subsequent cycles.

The feedback may be to the developers from the results of their own implementation and testing phase or to them from the user community working with a released version of the system.

3. The development team can gradually reapply their growing maturity with their tools— they are not required to apply all the best (and most complex) language/tool features from the beginning, or alternatively, only apply unskillful techniques to portion of the system.

4. Requirements can be adjusted to match changing business needs, as the project proceeds.

## Event-Driven Development Approach

Another new feature of modern programming is the emergence of event-driven systems. In older programs, the programmer was responsible for building complete, sequential programs that defined and controlled every step taken by the user. Modern window-based software does not follow a sequential process. Instead, actions by users generate events. The programs respond to these events and alter data or offer additional choices. Typical events include mouse clicks pointing to items on the screen, keystrokes, and changes to values, or transmissions from other systems.

Users will be asked to help identify important objects along with their attributes and functions. Users will also determine specific events and rules. In any business application, the underlying business rules define interrelationships between the objects. As certain events occur, data elements will be changed or added.

*Dr. Jennifer W. Gilmore, PhD*

## End-User Development Approach

The term end-user development simply means that users do all of the development work themselves. In many ways, it resembles prototyping, except that users (instead of analysts from the MIS department) create and modify the prototypes. Clearly the main advantage is that users get what they want without waiting for a MIS team to finish its other work and without the difficulty of trying to describe the business problems to someone else.

Two basic reasons explain why end-user development is increasingly popular. First, most MIS organizations are facing a two- or three-year backlog of projects. That means that if a new project is brought to MIS, the designers will not even start on it for at least two years (unless the project is given up for some other project). In fact, with the year-2000 changes, many MIDS departments simply gave up on other modifications. The second reason is that software tools are getting more powerful and easier to use at the same time. Today it is possible for users to create systems with a spreadsheet in a few hours that 10 years ago would have taken MIS programmers a month to build with third-generation languages. As tools become more powerful and more integrated, it becomes possible to create even more complex systems. Today, with windowing software users can build systems that share data with many users across the corporate networks—simply by pointing to items with a mouse. The advantages of end-user development are similar to those in prototyping. In particular, users get what they want, and they get working systems sooner.

## Potential Problems of End-User Development

The potential problems of end-user development are not always easy to see. Most of them arise from the fact that users generally lack the training and experience of MIS analysts and programmers. For instance, systems produced by end users tend to be written for only one person to use. They are oriented to working on stand-alone personal computers. The systems are often customized to fit the needs of the original users. Additionally, most users do not write documentation, so others will have difficulty using the products. Because of lack of training, users rarely perform as much testing, as they should. The systems lack security controls and are hard to modify.

Other problems stem from the bottom-up approach inherent in end-user development. People in different areas of the company will wind up working on the same problem, when it could have been solved once by MIS. Data tends to be scattered throughout the company, making it hard to share and wasting space. Not following standards generates incompatibilities among systems, making it difficult to combine systems created by different departments or by people within the same department.

End users are limited by the capabilities of commercial software. The initial systems may work fine, but as the company grows and changes, the commercial software might be unable to support the necessary changes. As a result, some users have created systems that produce incorrect answers, take too long to run, or lose data.

The last, and possibly most important, complication is that end-user development takes time away from the user's job. Some users spend months creating and modifying systems that might have been created by MIS programmers in a fraction of the time. One of the reasons for creating a MIS department is to gain efficiency from using specialists. If users are spending too much time creating and revising their own application, the company needs to consider hiring more MIS personnel.

## User-Interface Design Approach

Gossain (1998) described another method of systems development called user-centered design performed in parallel with object-oriented systems design as follows:

User-centered design is an approach to system design that involves users early in the development life cycle. This user involvement during design is intended to lead to systems that (1) are easier for experienced users to learn (2) are easier for new users to learn (3) allow for improved user productivity and (4) are based on user usage patterns.

The process outlined by Gossain involves users from the early scoping-analysis activity through to the delivery of the system, where the final stage is User Acceptance.

During user-interface design, the users are led through workshop sessions during which their tasks are explored and documented. Some business-process modeling may already have taken place during earlier phases, or prior to the high-level scope analysis, but now the focus is on user tasks and modeling these tasks. These work-products are used to provide input to the rapid object modeling activity being conducted in parallel.

These models of user processes are then used to work with users to identify the interaction they will have with the system through the screens of the user interface. Screen design is conducted on whiteboards and flip charts and the screen flow is also outlined in order to link the screens together for each task. Once the users have provided their input, these paper drawings can be used to create screens in the chosen user-interface development tool. It is common that at this point the wishes of the users may have to be tempered somewhat in order to satisfy usability requirements and for the interface to conform to standard user-interface guidelines.

Once the screens have been designed they are shown to the users, who are led through their tasks with the screen designs and screen flow; at this point changes can be made. Sometimes, wholesale changes are required after users see their wishes transformed into real screens.

Once the screens have been agreed on in principle, screen specifications can be created, identifying fields and their contents, and the events generated from each screen with the appropriate data. This event information is used to drive the object-interaction modeling. It is important to note that the user-interface design activity is not complete until users have 'signed off' on all screens and are happy with their look and with the screen flow. Principal work-products from this stage are (1) user task models based on business-process models, (2) screen specifications, and (3) screen flows to support business processes and use cases.

Larman (1997) in a similar vein to Gossain (1998) stated that typically, the user interface is the visible realization of the system. It is what the non-developers (managers, clients) see and viscerally think of as the system. People are visual oriented. Usually, an appealing user interface will provide client satisfaction, even when there is no significant functionality to go with it. Therefore, getting an early start on an alluring user interface is helpful. The downside is that clients, seeing the interface, think the system is nearly complete, when in fact it is hardly begun.

## Expert System Development Approach

McNurlin and Sprague, Jr. (1998) stated that the field of expert systems is no longer limited to laboratories and experiments. Many expert systems have been put in production, either through major expert system projects or by encouraging end users to write smaller knowledge-based systems. Since both types yield significant payoffs, it is believed IS management should follow a balanced approach to building these systems—encourage many small systems and carefully develop large systems. Expert systems, once viewed as an arcane use of IT, has become just another tool in the IS toolkit.

Turban and Aronson (1998) explained the following about building expert systems:

An expert system is basically computer software, so its development follows a software development process. The goal of such a process is to maximize the probability of developing viable, sustainable software within cost limitations and on schedule. The main functions of a model of this process are to determine the order of the steps (or tasks) involved in the software development and to establish the transition criteria for progressing from one stage to another. Many such models have been proposed; notable is the waterfall modes for a system development life cycle (SDLC).

When expert systems are constructed, most of the software development tasks are performed. The nature of the specific application determines which tasks are going to be performed, in which order, and to what depth. For example, a large-scale ES will be developed according to a complex life-cycle process, whereas a small-scale system for end-users will include only a few of the tasks.

The various tasks encountered in building expert systems are organized into six phases. However, the process is not linear; rather, some tasks are performed simultaneously and a return to previous tasks or even phases happens often.

Turban and Aronson (1998) provided details of these six phases as described below:

Phase I: Project Initialization

Finding an appropriate project for ES is not an easy task. Dozens of factors must be considered, and many ES projects fail because of poor up-front analysis. The typical tasks of this phase as described above are interrelated and might not follow any specific sequence. The details of the major tasks are discussed next.

## Problem Definition and Need Assessment

A clear definition of the problem simplifies the remaining tasks significantly and helps generate a productive program. Defining the problem is a matter of answering some basic questions. Whatever the problem or need, write a clear statement of it and provide as much supporting information as possible.

Many expert systems are used to improve an aspect of poor job performance. For example, an employee may not be achieving a desired quantity or quality of work within time or cost constraints. Problems such as this can often be traced to a lack of knowledge. The employee must either possess the knowledge or have access to it, whether in the form of an expert person or an expert system. The best way to understand the problem or need is to conduct a kind of formal study called a need assessment.

Assessing the need to solve the problem is related to several other activities such as cost-benefit analysis and justification and to finding the system requirements.

## Evaluation of Alternative Solutions

Before a major ES development program can be started, alternative solution approaches to the problem should be considered. Lack of knowledge is a deficiency that can be solved in ways other than with expert systems. Some examples are considered below.

If the problem is knowledge related, then usually someone must have the desired knowledge. One approach may simply be to make an existing or new expert accessible to those needing the expertise. This is a viable approach when the need for expertise is infrequent.

Another solution is to provide education and training in the desired subject matter for those who need it. Courses, seminars, and related materials may be much less expensive to develop than an expert system. In many cases, education and training can be provided at low cost via the Internet or Intranets.

An alternative to additional education and training is to package the knowledge and related information into printed documentation. Of course, it takes time to develop manuals and job aids. However, they are often fairly inexpensive and a great deal easier to create than an expert system.

A computer solution may still be best for a problem, but an expert system may not be completely appropriate. Once the problem has been defined, the possibility of using standard software packages should be examined. For example, a popular spreadsheet or database management system may work when the task involves model manipulation or database access.

There is a growing amount of knowledge that is disseminated, for a fee or for free, on the Internet. Such knowledge can be a substitute for building an expert system. The knowledge can be available in an ES and can be combined with online training.

### Verification of an Expert System Approach

The fact that other alternatives are not appropriate for solving a problem does not mean that an expert system is necessary. A framework for determining problem fit with an ES approach is proposed. According to this framework, a three-part study should be conducted: requirements, justification, and appropriateness.

The following 12 requirements are all necessary to make ES development successful:

1. The task does not require common sense.

2. The task requires only cognitive, not physical, skills.

3. There is at least one genuine expert who is willing to cooperate.

4. Experts involved can articulate their methods of problem solving.

5. Experts involved can agree on the knowledge and the solution approach to the problem.

6. The task is not too difficult.

7. The task is well understood and is defined clearly.

8. The task definition is fairly stable.

9. Conventional (algorithmic) computer solution techniques are not satisfactory.

10. Incorrect or nonoptimal results generated by the ES can be tolerated.

11. Data and test cases are available.

12. The task's vocabulary has no more than a couple of hundred concepts.

## Justification for ES Development

Like any other information system, an expert system must be justified. Of the following eight factors, at least one must be present to justify developing an ES:

1. The solution to the problem has a high payoff.

2. The ES can preserve scarce human expertise, so it will not be lost.

3. Expertise is needed in many locations.

4. Expertise is needed in hostile or hazardous environments.

5. The expertise improves performance or quality.

6. The system can be used for training.

7. The ES solution can be derived faster than that which a human can provide.

8. The ES is more consistent or accurate than a human.

The derived benefits in one or more of these areas must be compared to the costs of developing the system. A preliminary justification is conducted in this phase, whereas a detailed analysis is performed in Phase II.

## Appropriateness of the ES

Three factors should be considered in determining when it is appropriate to develop an ES:

1. Nature of the problem: The problem should have a symbolic structure and heuristics should be available for its solution. In addition, it is desirable that the task be decomposable.

2. Complexity of the task: The task should be neither too easy nor too difficult for a human expert.

3. Scope of the problem: The problem should be a manageable size; it also should have some practical value.

## Managerial Issues

ES projects do not start by themselves. Sometimes they start because there is an acute need. In many cases they start because someone in the organization believes in Artificial Intelligence (AI) technologies and is willing to support an ES project. A project may also start as soon as a decision is mode to look for an appropriate project. A project may start because the company has decided to follow a competitor's lead, or because one of the company's employees did an ES project as a student. Who starts a project is obviously an important issue, but there are several other managerial issues that must be considered when an ES (or any other AI project) is launched:

- Selling the project: All interested parties and especially top management must be convinced of the project's value.

- Identifying a champion: Someone in top management must sponsor the project strongly.

- Level of top management support: This can directly influence the level of success of the system.

- End-user involvement, support, and training: End-users must be involved early in the project. Their proper support and training are critical.

- Availability of financing.

- Availability of other resources.

- Legal and other potential constraints.

Phase II: System Analysis and Design

Once the concept of the project has been approved, a detailed system analysis must be conducted to estimate the functionality of the system. Many tasks are performed in this phase.

Conceptual Design

A conceptual design of an ES is similar to an architectural sketch of a house. It gives you a general idea of what the system will look like and how it is going to solve the problem. The design shows the general capabilities of the system, and interfaces with other computer-based information systems, the areas of risk, the required resources, anticipated cash flow, the composition of the team, and any other information that is necessary for detailed design later. Once the conceptual design is completed, it is necessary to determine the development strategy.

## Development Strategy and Methodology

There are several general classes of AI development strategies: do it yourself, hire an outside developer, enter into a joint venture, merge, and attach on all fronts. Although some organizations use a single strategy, others use several.

## Selecting an Expert

### Experts

Human experts possess knowledge that is much more complex than what we find in documented sources. It is based on experience, and in many cases can be expressed in terms of heuristics.

### Selection

Expert systems use both documented sources and human experts as sources of knowledge. The more human expertise is needed, the longer and more complicated the acquisitions process becomes. In general, management should use some incentives to influence experts so that they will cooperate fully with the knowledge engineer. Once the experts are identified, the builder turns to software considerations.

## Software Classification: Technology Levels

It is helpful to classify ES software into five technology levels: languages, support tools, shells, hybrid systems, and ES applications (specific ES). Roughly speaking, the specific application can be constructed with one or more shells, support tools, hybrid systems, or languages. Shells and hybrid systems can be constructed with languages or support tools, and support tools are constructed with languages. The higher the level of the software, the less programming is required. The trade-off is that the higher the level, the less flexible is the software. Generally speaking, the use of higher levels of software enables quicker programming (even by end-users). On the other hand, complex and nonstandard applications must be built with lower levels of software.

With the increased complexity of problems for which ES are constructed there is a trend to increase the use of frames and other object-oriented implementations. Based on the desire to disseminate expertise organization-wide or to customers, any place and any time, many vendors have expanded their ES shells, toolkits, and languages to provide an ES interface directly through Web browsers such as Netscape Navigator or Microsoft Internet Explorer. Expert systems are programmed in COBOL, Pascal, C, C+, and C++.

Feasibility Study

The feasibility study for an ES is similar in structure to a feasibility study for any information system. The larger the system, the more formal the steps must be because approval by top management is required. Each ES project requires an investment of resources, which can be viewed as the cost of the system, in exchange for some expected benefit. Finding an appropriate ES application is a critical success factor. A major task in such a search is the cost-benefit analysis.

Phase III: Rapid Prototyping and a Demonstration Prototype

Prototyping has been crucial to the development of many ES. A prototype in ES is a small-scale system. It includes representation of the knowledge captured in a manner that enables quick inference and the creation of the major components of an ES on a rudimentary basis. The prototype helps the builder decide on the structure of the knowledge base before spending a great amount of time on building more rules. Rapid prototyping is essential in large systems, because the cost of a poorly structured and then not used ES can be very high.

The process of rapid prototyping is as follows. The design of a small system is started with. The designer determines what aspect (or segment) to prototype, how many rules to use in the first cut, and so on. Then the knowledge is acquired for the first cut and presented in the ES. Next, a test is conducted. The test can be done using historical or hypothetical cases. The expert is asked to judge the results. The knowledge representation methods and the software and hardware effectiveness are also checked. A potential user may be invited to test the system.

The results are then analyzed by the knowledge engineer, and if improvement is needed the system is redesigned. Usually the system goes through several iterations with appropriate refinements. The process continues until the system is ready for a formal demonstration. Once the system is demonstrated, it is tested again and improved. This process continues until the final (complete) prototype is ready.

One advantage of rule-based systems, especially when combined with frames is that they are modular. Thus small subdivisions can be constructed of larger systems one step at a time and can be tested. The prototyping phase can be short and simple or it can take several months and be fairly complex. The lessons learned during rapid prototyping are incorporated into the final design. Also, this is the time when another go/no-go decision is made. If the decision is 'go,' system development begins.

Phase IV: System Development

Once the initial prototype is ready and management is satisfied, system development begins. Obviously, a plan must be made about how to continue. At this stage, the development strategy may be changed (for example, a consultant may be hired). The detailed design is also likely to be changed, and so are other elements of the plan.

Depending on the nature of the system—its size, the amount and type of required interfaces with other systems, the dynamics of the knowledge, and the development strategy—one or both of the following approaches for system development will be used:

1. Continue with prototyping

2. Use the structured life-cycle approach

System development can be a lengthy and complex process. In this phase, the knowledge base is built and continuous testing, reviews, and improvements are carried out. Other activities include creation of interfaces (for example, to databases, documents, multimedia objects, hypermedia, and even the Web), creating and testing the users interface and so on.

Phase V: Implementation

The process of implementing an ES can be long and complex, like the implementation of any software project.

Acceptance by the User

Acceptance depends on behavioral and psychological considerations, as well as on quality and ease of use. It is important that the development of specific ES be communicated as widely as possible to foster a climate of acceptance among the people who will use the system. Behavioral aspects play a major role in dealing with the user.

Installation Approaches

The expert system is ready to be field tested when it reaches a minimal level of stability and quality. In rule-based systems, this may occur when the system can handle 75 percent of the cases and exhibit less than a 5 percent error rate. The system can be installed in parallel with a human expert for a short test period.

Demonstration

Demonstrating the fully operational system to the user community is important. Viewers can become believers.

Mode of Deployment

Several deployment modes for ES may be considered. The final system could be delivered to users as a turnkey, standalone system. It could be operated as a separate entity but integrated into the users' environment, it could be embedded into another system, or it could be run as a service, with the users' requests and data accessed remotely and results delivered to the users. The Web is a convenient way of integrating an ES into the users' environment.

Orientation and Training

Sufficient and high-quality user training is an important factor in user acceptance of ES. Depending on the mode of deployment, the builders must plan appropriate orientation and training. If the users are assigned maintenance responsibilities, the training may be fairly extensive.

Security

Security is a heightened concern in ES that may contain the accumulated proprietary knowledge of a firm. Communicating and distributing the end product, protecting the software, and at the same time providing an environment that does not constrain authorized users is a substantial practical problem.

Documentation

Implementation of ES must include appropriate documentation. The planned documentation to accompany the system might consist of printed manuals, online documentation, or both. There may be different sets of documentation for system maintainers, operators, and users. Specific recommendations for documentation include the following: the system overview, a technical description, the high-level map of the whole problem, and maps of the individual tasks. Also, an index of all items within a knowledge base that depend on actions outside that knowledge base, a record of all computer files used, and printed versions and backups of all computer files used. For user documentation the following should be included: an introductory brochure, the system overview, a brief user guide, and a means to encourage users to provide feedback on the system.

Integration and Field Testing

If the expert system stands alone, it can now be field-tested. A system that must be integrated is physically added to the existing computer-based information system before testing in the field is conducted. Field-testing is extremely important because conditions in the field may differ from those at the developer's lab.

## Phase VI: Post-implementation

Several activities are performed once the system is distributed to users. The most important of the activities are system operation, maintenance, upgrading and expansion, and evaluation.

Operation

If the expert system is to be delivered as a service, a system operations group (or several groups if there are several sites) should be formed and trained. If the system is to be a product run by users, an operator-training group may need to be formed, and consideration should be given to providing help for user-operators with problems. If the system is embedded into another system, the operators of the other system should be trained in any new operation procedures required.

Maintenance

Because an expert system evolves, it is never really done. Thus, it is important to plan the maintenance. It must be revised on a regular basis with regard to the applicability of the rules, the integrity and quality of the data feeds, the use of the interlinked databases, and so on. Because experts are constantly training themselves on new situations, or reorganizing their knowledge to account for unencountered situations, an ES must be adjusted for these cases. In addition, software and hardware bugs must be fixed as found and the system must be upgraded to run on new software (due to upgrades from the vendor) and hardware platforms. Just as in the system development life cycle, there is a maintenance cycle, and a long-term maintenance team must be formed and trained to perform the above tasks.

If the expert system is embedded in another system, some thought should be given to whether one maintenance group will serve the overall system or whether the expert system will be maintained separately. If separate maintenance is chosen, procedures for coordinating the two maintenance groups should be formed.

Expansion (Upgrading)

Expert systems continuously evolve and therefore continuously expand. All new knowledge must be added, and new features and capabilities must be added, as they become available. The upgrading tasks generally fall under the jurisdiction of the maintenance team. However, the original builder or even a vendor can also carry out some expansion. Some systems can automatically capture problem areas to be addressed in the next version.

Evaluation

Expert systems need evaluation periodically (for example, every 6 or 12 months). In such an evaluation, questions such as these should be answered:

1. What is the actual cost of maintaining the system as compared to the actual benefits?

2. Is the maintenance provided sufficient to keep the knowledge up to date so that system accuracy remains high?

3. Is the system accessible to all users?

4. Is acceptance by users increasing?

If the users have been encouraged to comment on the ES, and the feedback process is easy, it is much easier to maintain a system because of continuous feedback. This also reduces the difficulty of the periodic evaluation.

## The Decision Support System Development Process

Turban and Aronson (1998) described the DSS development process as follows:

Because of the semi-structured or unstructured nature of problems addressed by decision support systems, managers' perceived needs for information may be unclear. Therefore, most DSS are developed by a prototyping process.

Phase A: Planning

Planning deals mainly with need assessment and problem diagnosis. Here the objectives and goals of the decision support system are defined. A crucial step in the planning effort is determining the key decisions to be supported by the DSS. For example, in a portfolio selection system, a key decision might be selecting the correct stocks for a customer's needs.

Phase B: Research

This phase involves the identification of a relevant approach for addressing user needs and available resources (hardware, software, vendors, systems, studies or related experiences in other organizations, and review of relevant research). Here, the DSS environment is closely checked.

Phase C: System Analysis and Conceptual Design

This phase includes the determination of the best construction approach and specific resources required to implement it, including technical, staff, financial, and organizational resources. It is basically a conceptual design followed by a feasibility study.

Phase D: Design

The detailed specifications of the system components, structure, and features are determined. The design can be divided into four parts corresponding to the major components of a DSS. Here one selects appropriate software or writes it. A major issue in the design effort is deciding which of the many commercially available software packages to use.

Phase E: Construction

A DSS can be constructed in different ways depending on the design philosophy and the tools being used. The construction is the technical implementation of the design. As the system is constructed, it is tested continuously and improved. In this phase, the DSS is connected, if needed, with appropriate Computer Based Information Systems (CBIS) and networks.

Phase F: Implementation

The implementation phase consists of the following tasks: testing, evaluation, demonstration, orientation, training, and deployment. Several of these tasks are performed simultaneously.

Phase G: Maintenance and Documentation

Maintenance involves planning for ongoing support of the system and its user community. Proper documentation for using and maintaining the system is also developed.

Phase H: Adaptation

Adaptation requires recycling through the earlier steps on a regular basis to respond to changing user needs. This process has several variations because of the many variations of DSS.

## Neural Network Application Development

Turban and Aronson (1998) described the development of neural network applications as follows:

Although the development process of Artificial Neural Networks (ANNs) is similar to the structured design methodologies of traditional computer-based information systems, some steps are unique to neural network applications or have additional considerations. In the process described here, it is assumed that preliminary steps of system development, such as determining information requirements and conducting feasibility analysis for the project, have been completed successfully. Such steps are generic to any information system.

The development process of an ANN application has nine steps:

In Step 1, the data to be used for training and testing of the network are collected. Important considerations are that the particular problem is amenable to neural network solution and that adequate data exist and can be obtained. In Step 2 training data must be identified, and a plan must be made for testing the performance of the network.

In Steps 3 and 4 a network architecture and a learning method are selected. The availability of a particular development tool or the capabilities of the development personnel may determine the type of neural network to be constructed. Also, certain problem types have demonstrated high success levels with certain configurations. Important considerations are the particular number of neurons and the number of layers.

Current neural network models have parameters that tune the network to the desired performance level. Part of the process in Step 5 is initialization of the network weights and parameters, followed by modification of the parameters as performance feedback is received. Often, the initial values are important for determining the efficiency and length of the training.

The next procedure, Step 6, transforms the application data into the type and format required by the neural network. This may mean writing software for preprocessing the data. Data storage and manipulation techniques and processes must be designed for conveniently and efficiently retraining the neural network when needed. Also, the way the application data are represented and ordered often determines the efficiency and possibly the accuracy of the results from the network.

In Steps 7 and 8, training and testing are conducted as an iterative process of presenting input and desired output data to the network. The network computes the actual outputs and adjusts the weights until the actual outputs match what is desired. The desired outputs and their relationships to input data are derived from historical data (a portion of data collected in Step 1).

In Step 9 in the process, a stable set of weights is obtained. Now the network can reproduce the desired outputs given inputs such as those in the training set. The network is ready for use as a standalone system or as part of another software system.

## Benefits of Neural Networks

Turban and Aronson (1998) described the benefits of neural networks as follows:

The value of neural network technology includes its usefulness for pattern recognition, learning, classification, generalization and abstraction, and the interpretation of incomplete and noisy inputs. A natural overlap with traditional AI applications thus occurs in the area of pattern recognition for character, speech, and visual recognition. Systems that learn are more natural interfaces to the real world than systems that must be programmed. Speed considerations point to the need to take advantage of parallel processing implementations.

Neural networks have the potential to provide some of the human characteristics of problem solving that are difficult to simulate using the logical, analytical techniques of expert systems and standard software technologies. For example, neural networks can analyze large quantities of data to establish patterns and characteristics in situations where rules are not known. Neural networks may be useful for financial applications such as measuring stock fluctuations for determining an appropriate portfolio mix. Likewise, neural networks can provide the human characteristic of making sense of incomplete or noisy data and can perform data mining, identifying patterns in large databases. These features have thus far proven too difficult for the symbolic/logical approach of traditional AI.

Neural networks have the following other benefits: Ability to tackle new kinds of problems, robustness, fast processing speed, and flexibility and ease of maintenance.

Thus, neural computing differs from traditional computing methods in many ways, and the application developer can exploit the differences. Neural networks can be applied in areas where data are multivariate, with a high degree of interdependence among attributes, data are noisy or incomplete, or many hypotheses are to be pursued in parallel, and high computational rates are required.

Beyond its role as an alternative, neural computing can be combined with conventional software to produce powerful hybrid systems. Such integrated systems can include database, expert system, neural network, and other technologies to produce computerized solutions to complex problems. ANN are especially suitable to complex decision support.

## Limitations of Neural Networks

According to Turban and Aronson (1998) neural networks have the following limitations:

In general, ANNs do not do well at tasks that are not done well be people. For example, arithmetic and data processing tasks are not suitable for ANNs, and are best accomplished by conventional computers. ANNs currently excel in the areas of classification and pattern recognition.

Most neural network systems lack explanation capabilities. Justifications for results are difficult to obtain because the connection weights do not usually have obvious interpretations. This is particularly true in pattern recognition, where it is very difficult or even impossible to explain the logic behind specific decisions. However, some ANN tools offer some explanations by analyzing the importance of the input data in relationship to the output. But when seemingly insignificant inputs (statistically speaking) are removed from the network and it is retrained, the predictions can become very bad. The limitations and expense of current parallel hardware technology restrict most applications to software simulations. With current technologies, training times can be excessive and tedious; thus, the need for frequent retraining may make a particular application impractical. Finally, neural computing usually requires large amounts of training and test data.

## The Development of Executive Information Systems

Sprague and Watson (1996) explained that building a successful EIS is challenging. A myriad of technical, organizational, and managerial issues must be addressed. Of utmost importance is creating an EIS that is easy to use. Consequently, system designers should pay careful attention to the design of the user interface for the system.

### The EIS User Interface

The user interface refers to how the user directs the operation of the system, for example, keyboard, mouse, or touch screen; question/answer, command language, or menus) and how the output is given to the user, for example, graphical, tabular or textual; color or monochrome; paper or online. For the system to be easy to use, the user must know how to make it work and what the output means. A user interface must be designed to make operating the system and interpreting the output as easy as possible.

Designing an EIS user interface is somewhat different from designing other information systems. Because of the nature of executive users, the system must be more than user friendly; it must be user intuitive, even user seductive. Another difference is the flexibility the system must have, because it is difficult to determine how a particular executive will use an EIS. Also, because of advances in hardware and software, system designers have many new options to choose from when implementing an EIS.

A successful EIS often benefits other users in addition to executives. For this reason, EIS also stands for 'everybody's information system'. These users are more likely to accept complex user interfaces than senior executives are and may be willing to trade off simplicity for flexibility. In many instances, however, the more complex applications created for lower organizational-level users are not given to executives.

## Design Guidelines

The following eight guidelines on designing an EIS interface should help developers successfully implement an EIS:

1. Involve executives in the design of the user interface.

2. Set standards for screen layout, format, and color.

3. Use of the system should be intuitive.

4. Use standard definitions of terms.

5. Design the main menu as a gateway to all computer use.

6. Design the system for ease of navigation.

7. Strive to make response time as fast as possible.

8. Expect preferences in user interfaces to change.

The following paragraphs describe these guidelines in more detail.

Involving Executives in the Design of the User Interface

Although user involvement in the systems development process is critical for all types of information systems, executive involvement in the design of an EIS user interface is especially important. Executives might have limited experience working directly with a computer, and if they do have some computer experience, the EIS will look and feel quite different from any other organizational information systems the executives might know. Designers should be prepared to show a variety of prototype screens and navigation approaches because the executive might have limited knowledge of what an EIS can actually do. Evaluating these prototypes is also likely to get apprehensive executives more committed to the EIS as they begin to see the system's potential. For this reason, it is important to involve all executive users in the process, not just the executive sponsor.

Prototyping Approaches

Early prototyping should be used to help decide on the basic look and feel of the system. Two fundamental approaches should be presented:

1. A full-screen interface with large buttons and icons

2. A multiple window interface with pull-down menus and dialog boxes.

The first approach might be less intimidating, but the second approach conforms to the popular interface design standards. The preferred look and feel should be used to finalize the development environment that will be used, for example, Windows, as some development environments might more easily accommodate one or the other type of system. Additionally, differences in preference for the look and feel reveal early in the development process the amount of individual tailoring for each executive that might be required.

Most of the commercial EIS products are designed to run in a Windows environment and to take advantage of the functionality of Windows. It is usually not safe to assume that executives have experience with Windows and can easily operate Windows-based applications. When users are not proficient with Windows, there are two options. The first is to train them in Windows and then in the application. The second is to design the EIS in such a way that the features of Windows are not used. Experience has shown that an application can fail, not because it is poor, but because it incorrectly assumes a proficiency in Windows.

Although rapid prototyping and extensive user feedback are quite important, the prototypes do not have to be computer-based. Paper screen mock-ups or storyboards can be quite effective because the executive can review the screens as time permits and consider the alternatives before providing feedback to the designer. Computer-based prototypes, however, are quite useful when showing the executive the potential of the technology and when exploring navigation approaches the executive might prefer.

Executives also must be involved in the design of the interface, because preferences for screen prototypes can provide clues about the importance of screen content and design. This aids a designer in uncovering additional information requirements. The relationships among importance of data, the level of detail desired, and the frequency of need for the information can help a designer understand the way an executive will actually use the EIS.

Because of the almost endless number of possible screens that can be provided, it is important for the designer to narrow the number down to the most important screens for each executive. This not only reduces development time and system overhead, but also makes it possible to provide a system that makes it easy for the executive to find the information that is actually needed.

Any later changes to the interface of an EIS should be discussed with its users. This is especially true when a designer considers deleting seldom-used screens. It is not easy to tell the value of a particular screen just by tracking usage. An executive may have looked at a particular screen only once, but that screen could have provided critical insight that day. Months later, the same screen might be needed once again when the same critical need arises.

Setting Standards for Screen Layout, Format, and Color

Currently available EIS software offers an array of screen design alternatives. Screens can display graphs, tables, and text in hundreds of formats and colors. Unfortunately, this cornucopia of choices can be detrimental. There is a temptation to use many of these alternatives to add sizzle to the screens, but yielding to this temptation can create displays that are confusing. Designers should carefully develop screen design standards that use only a few layouts, formats, and colors.

Advanced Capabilities

A few EISs allow voice commentaries to be associated with screens. This is an appealing feature because executives are used to receiving information verbally and voice is richer for communications than printed words. Voice annotations to screens are examples of the multimedia enhancements to EISs. Other possibilities such as video and personal teleconferencing have good potential.

Using the System Should be Intuitive

Ideally, an executive should be able to use an EIS without training. At the most, no more than 15 minutes of instruction should be required to teach how to use the basic information-retrieval capabilities. Systems more complex than this are unlikely to be used.

Most successful EISs are operated by point-and-click technology. By picking from among menus, icons, or buttons, an executive navigates through the system to a desired capability, for example, e-mail or information. Experience with decision support systems has shown that most executives will not use a command language with a verb-noun syntax because it is too time-consuming to use and difficult to learn and remember.

User Documentation

Systems developers are typically expected to write user documentation for new applications. However, this is usually unnecessary or inappropriate for EISs. The system should be sufficiently intuitive that instruction manuals are not needed. Even more so than with other types of users, executives do not read documentation. If an executive is having a problem using the system, it is best if the user calls the EIS support staff to correct the difficulty.

Users may request documentation, and in this case, it should be provided, either within the system and or as hard copy. Ideally, the instructions should fit on a single page or a few screens.

Using Standard Definitions of Terms

Most organizations have data dictionaries that include definitions for the data elements used in transaction processing applications. There are other terms that are widely used throughout organizations and are very important to EISs that are not as precisely defined. Everyone in a company uses these words and has a general understanding of their meaning but slight differences exist and can cause misunderstandings.

Designing the Main Menu as a Gateway

Most organizations have a variety of applications designed to support executives: e-mail, electronic filing, decision support, and access to external news and stock prices. It is common for many of them to require their own access procedures and passwords. This requirement, and the resulting difficulty and inconvenience, discourages hands-on computer use. The development of an EIS provides an excellent opportunity to deliver all of these capabilities in a single, integrated system. An EIS provides the logical and physical umbrella under which all of the executives' computer applications are placed.

A number of EISs use their main menus to display all information and applications available through picks (i.e., menus, icons, or buttons). The kinds of information usually provide one set of options. For example, there may be screen picks for financial, production, marketing, and human resources information. Separate picks may exist on the basis of products, geographical location, and organizational units, for example, corporate or division. The choices reflect the information contained in the EIS and how it is organized. Lower-level menus let users move to specific information desired within a general category.

Access to these applications should be transparent to the user and not require any additional log-on procedures or passwords; these activities should be handled automatically by the system.

Designing the System for Ease of Navigation

Vendors' demos often show executives moving easily through a system, looking at current status information and drilling down to more detailed information when a problem or item of interest is identified. This scenario is possible in practice, but only if careful attention is given to navigation issues early in the system's design.

Navigation problems may be masked when there are few screens in the system. As the number of screens grows, as they inevitably do, users find it more difficult and time-consuming to move through a system. For example, and executive is looking at financial information and wants to move to operational production data. In a poorly deigned system, the user will have to back out of the financial application, screen by screen, until the main menu is reached, and then enter the production application, and move through screens to the desired information.

The starting point in designing navigation for an EIS is understanding the mental models that executives have of the organization. If the structure of the information does not match their mental models, users will have a difficult time finding the information they want. For example, do executives look at the firm in terms of geographical location, products, function areas, or divisions? Each view of the organization may call for a pick on the main menu and a set of related screens.

A complicating factor is when one or a few executives have unique mental models. During the development of one EIS at a hospital, designers found that the director of nursing wanted information structured much differently than other users. Her view of the hospital could be accommodated but required custom designing the system for her use. The decision of whether to do this was a business rather than a technical one.

Navigation Features

There are features that can be included in an EIS to make navigation easier. Some systems have a screen that shows where the user is in the system. Often, users get lost and are uncertain about how to move elsewhere, short of turning off the system and starting over. Another feature is to have a home key or pick that takes the user directly back to the main menu. Some systems provide a retrace capability that allows users to easily backtrack to screens viewed previously. Another helpful feature is to include a pick on the main menu that takes the user to a screen that lists the user's most popular screens. From this screen, a user can go directly to any screen on a personalized menu. Also, a single menu can be created that provides direct access to a large number of screens.

Response Time as Fast as Possible

When incorporating text and graphics, internal and external data, hundreds of individually tailored screens and views, and multiple navigational paths through the system, EIS developers must continually monitor the response time of the system. Executives are intolerant of slow response times.

Response time problems can be anticipated when the EIS must dynamically build a screen each time it is requested by searching corporate databases. Response time can be much faster if screens are static and updated each night, though designers must evaluate the trade-off between timeliness of data on the screens and response time. Response time can also be affected by the bandwidth of today's networks.

When Speed Counts

Generally, executives expect very fast response times when flipping through their usual set of screens each morning. One EIS developer suggested thinking of the maximum acceptable time to move from screen to screen as the time it takes the executive to turn a page of *The Wall Street Journal*.

Executives can usually tolerate a slow response to ad hoc queries. When an executive is used to waiting several days for the staff to gather information for a specific question, several minutes may be an acceptable wait for directly retrieving the same information through the EIS. The differences between predefined screens and ad hoc query screens should be made clear to the executive, however. In either case, when any system function takes more than a few seconds, a message should always provide feedback that the system is processing the executive's request.

User Preferences May Change

Almost all aspects of an EIS, including the user interface, change in time. Often, organizations developing an EIS order touch screens for technophobic executives. These users quickly discover the disadvantages of touch screens and also find that using a mouse is easy after a little practice. Although touch screens may help sell the idea of an EIS to some executives, these executives will probably prefer mice eventually.

As an EIS evolves, the number of its users usually increases. Quite possibly, training given to first-time users will have to change. For example, more time may have to be spent discussing how to interpret the information presented on the most complex screens. Another approach is to include less complex screens in the system.

## Combining Methodologies

Post and Anderson (2000) stated that in actual practice, most companies use a combination of the various methods to create systems. Projects can be combined so that projects gain the benefits of each method. For example, large projects that require control and oversight benefit from using SDLC and other MIS-controlled methods. Even in these projects, prototyping and joint application development (JAD) are often used for reports and input screens. An MIS analyst will meet with users and quickly build sample screens and reports using CASE tools and personal computer-based software. Users alter the designs and they are adjusted until the users are satisfied. Then the prototype designs are turned over to the SDLC team to be programmed into the main system.

Some projects might begin as end-user systems to solve problems within one department. As the application expands and more users become involved, a RAD (rapid application development) or SDLC team might be called in to standardize the processes, convert the data, and build a shared application for the entire company.

Similarly, a company might use SDLC techniques (especially standards and testing) to create an initial set of objects. Once these objects are created, a team can use prototyping and RAD techniques to build new systems. Likewise, companies will purchase prewritten components as much as possible. It is always cheaper to buy components that already exist. These components can range from application software such as database management systems to object libraries that can be customized for each company.

Any modern company with a large enough MIS staff is continuously experimenting with all of these design methods. Creating software is an expensive, time-consuming process. Firms are constantly searching for techniques and methodologies that will improve the process and lead to better systems.

McLeod (1995) stated that for small-scale systems, prototyping could replace the system development life cycle. However, for large-scale systems or those that affect large organizational units, prototyping is incorporated into the SDLC. Prototyping can be used in all of the life cycle phases.

## Domain Experts

Gossain (1998) explained that the purpose of analysis is to understand a problem domain. Naturally, a domain expert can be a very useful ally during this process. A domain expert is a person that has experience working within a specific area of concern. Ideally, the person is somewhat introspective, too (someone who has spent some time thinking about what he or she does, fitting that work within an overall scope or larger perspective).

It is important to make the distinction between a domain expert and a user. For example, experts who may assist in understanding retail systems where point-of-sale terminals are the user tools are not usually the people who will eventually operate those terminals. The domain expert and the user can sometimes be the same person, however.

Domain experts can sometimes be instrumental in uncovering requirements that may be documented elsewhere, and can also provide the insight needed to fully understand the more complex aspects of a domain. In doing so, domain experts are often responsible for identifying key abstractions and relationships of the problem space that can allow an order-of-magnitude improvement in the expression of the domain concepts. Above all, domain experts can validate the object models.

There are a number of ways in which experts can be involved in the modeling process:

Interviews: These can be particularly useful if experts are not directly involved in conducting the analysis. A number of short directed sessions of 1-2 hours' duration are most profitable. The modeler should be well prepared before the interview with a ready set of questions.

Reviewing models: Experts can make excellent reviewers of models, as they will be best able to understand if the models being constructed best describe the problem domain. Just sending the models to an expert to review as if they were reading a book will not be sufficient, however. Early on, the expert will need to be walked through the models and every attribute and relationship will need to be explained and justified.

Participation in the analysis: Full-time participation of experts during the modeling process is the ideal. Expert participation ranges from occasional involvement to full-fledged team membership, however. When involving experts, seeding each team with an expert can often be the best way to gain maximum leverage out of his or her involvement. In such a situation, however, the experts should be fully trained in the approach being used.

There are a number of avenues through which to explore a problem domain. It is often best to use the path that the experts are most comfortable with, rather than forcing them to try to think the same way the analysts do.

## Types of Management Support Systems

Turban and Aronson (1998) described the following types of management support systems:

Decision Support Systems

A computerized decision support system may be needed for various reasons, such as, speedy communications, overcoming cognitive limits in processing and storage, and because of cognitive limits. Other reasons include cost reduction, technical support, quality support, and competitive edge: business process reengineering and empowerment.

Decision support systems couple the intellectual resources of individuals with the capabilities of the computer to improve the quality of decisions. It is a computerized support system for management decision-makers that deal with semi-structured problems. DSS is sometimes used as an umbrella term to describe any and every computerized system used to support decision making in an organization. The perceived benefits of DSS are higher decision quality, improved communication, cost reduction, increased productivity, time savings, and improved customer and employee satisfaction.

Another reason for the development of DSS is the end-user computing movement. End-users are not programmers, so they require easy-to-use construction tools and procedures. These are provided by DSS.

Group Decision Support Systems

Groups make many major decisions in organizations. Getting a group together in one place and at one time can be difficult and expensive. Furthermore, traditional group meetings can take a long time and the resulting decisions may be mediocre.

Attempts to improve the work of groups with the aid of information technology appear under several names, such as, groupware, electronic meeting systems, collaborative systems, and group decision support systems.

Executive Information (Support) Systems

Executive information systems (EIS) are developed primarily for the following objectives:

1. Provide an organizational view of operations

2. Serve the information needs of executives and other managers

3. Provide an extremely user-friendly interface that meets individual decision styles

4. Provide timely and effective tracking and control

5. Provide quick access to detailed information behind text, numbers, or graphics

6. Filter, compress, and track critical data and information

7. Identify problems (opportunities)

Executive information systems, which started in the mid-1980s in large corporations, have spread around the globe, have become affordable to smaller companies, and are serving many managers as enterprise-wide systems.

Expert Systems

When an organization has a complex decision to make or problem to solve, it often turns to experts for advice. These experts have specific knowledge and experience in the problem area. They are aware of the alternatives, the chances of success, and the benefits and costs the business may incur. Companies engage experts for advice on such matters as which equipment to buy, mergers and acquisitions, and advertising strategy. The more unstructured the situation, the more specialized (and expensive) is the advice. Expert systems attempt to mimic human experts.

Typically, an expert system (ES) is a decision-making or problem-solving computer package that can reach a level of performance comparable to—or even exceeding—that of a human expert in some specialized and usually narrow problem area.

The basic idea behind an ES, which is an applied artificial intelligence technology, is simple. Expertise is transferred from the expert to a computer. This knowledge is then stored in the computer and users call on the computer for specific advice as needed. The ES can make inferences and arrive at a specific conclusion. Then, like a human consultant, it advises the non-experts and explains, if necessary, the logic behind the advice. Expert systems are used today in thousands of organizations and support many tasks. Expert systems are often integrated with other information technologies.

Artificial Neural Networks

The application of the previous technologies was based on the use of explicit data, information, or knowledge, which was stored in a computer and manipulated as needed. However, in the complex real world we may not have explicit data, information, or knowledge. Thus, people must make decisions that are based on partial, incomplete, or inexact information. Such conditions are created, for example, in rapidly changing environments. Decision-makers use their experiences to handle these situations; that is, they recall experiences and learn from their experiences what to do with new similar situations for which exact replicas are unavailable.

In all the previous technologies, there was no element of learning by the computer. A technology that attempts to close this gap is called neural computing or artificial neural networks. The technology, which uses a pattern recognition approach, has been used successfully in many business applications.

Hybrid Support Systems

The objective of a computer-based information system (CBIS), regardless of its name or nature, is to assist management in solving management or organizational problems faster and better than what can be done without computers. To attain this objective, they may use one or more information technologies. Significant benefits can be found when integrating systems. When a problem solver uses several tools the different tools can be used in different ways to solve the problem.

## Reasons for Success of Systems

Post and Anderson (2000) listed the top five main reasons for the success of systems development as being the following: (1) User involvement, (2) Executive management support, (3) Clear requirements, (4) Proper planning, and (5) Realistic expectations.

## Reasons for Failure of Systems

Post and Anderson (2000) listed the five top reasons for failure of systems development as being the following: (1) Lack of user input, (2) Incomplete requirements, (3) Changing requirements and specifications, (4) Lack of executive support, and (5) Lack of technical skills.

## Summary

This chapter presented the review of the literature related to the topic of this researcher's study, different approaches of systems development. Chapter three presents and describes the methodology to be used for this study which is the case study method.

# METHODOLOGY

## Chapter 3

## Introduction

This chapter describes the methodology and research design being used in this study. It is a case study methodology describing two different approaches to systems development completed by this researcher for the development and successful implementation of two business software computer systems.

## Approach

The approach for the method of research for this dissertation is a case study approach. This researcher will be describing the steps in the development of two system development projects from beginning to end. The highlights of the projects will be covered detailing the two types of system development methods used. The first project used the System Development Life Cycle and the second project used Prototyping. The different steps in these two types of system development approaches will be mentioned and the major strides and pitfalls encountered will be described. For the System Development Life Cycle project how the Year 2000 COBOL Coding Problem was solved for this system will also be covered. Throughout the descriptions, reasons for the success of these two projects will be encountered and revealed.

**Data Gathering Method**

The data gathering method is mainly from the personal experience of this researcher with the two projects, as a project leader, senior systems analyst, and project manager. The written documentation of the two projects is also a large source of data for this research. From the beginning phases of feasibility study to acceptance and use of the systems by the users, data will be presented from the two system projects. Interviews with the users and managers of the implemented projects will also be included.

**Database of Study**

The database of the study will mainly be from the written documentation of the two projects from the beginning stages to the end. Other data are from the user sites where the actual human subjects and computer software and hardware reside.

**Analysis of Data**

The analysis of the data will be done based on the guidelines for the two methods of system development as presented in the literature review for the System Development Life Cycle and Prototyping methods. The different steps and junctions of the actual data as they pertain to the two systems that were developed will be highlighted. Additional steps used in the development of the two systems that are outside of the literature will also be mentioned. Statistical analyses and measurements will also be described to gauge the actual degree of success that was gained from the development of the two systems.

## Validity of Data

The data are valid in that they are data from two actual systems development projects that were successfully implemented in New York City at the Human Resources Administration Department. These data are heavily documented and could be readily checked and verified.

## Originality and Limitation of Data

The originality of the data is that they are unique to only these two systems development projects for the City of New York, Human Resources Administration Department. The limitation of the data is that the data are based on these two projects which spanned from 1990 to 1998. The Systems Development Life Cycle project lasted from 1990 to 1997 and the Prototyping project lasted from 1996 to 1998.

## Uniqueness and Limitations of the Method

The uniqueness and limitations of the method are that this method is unique to only the development of these two systems development projects. This method is limited to only these two projects although many of the steps in the literature for methods used for the System Development Life Cycle method and the Prototyping method are used and are included.

## Summary

The research method used for this study was described in this chapter. Chapter four will present the description of the actual two case studies of the development of the two systems development projects using the System Development Life Cycle method and the Prototyping method. Chapter five will present the highlights of these two projects and the reasons for the success of these two projects and the conclusions, implications, and recommendations for further research.

# PRESENTATION OF THE DATA

# Chapter 4

## Introduction

This chapter describes the two case studies and the steps in the development of the two successfully implemented systems. The first system using the systems development life cycle method of development is described first. The second system using the prototyping method of development is described second. The first system is called the Contract Agency Monthly Financial Report System (CAMFR) and the Front End Redesign was the first phase of this system development started in 1990. Other phases of development completed through the years are also described culminating in the last phase of development in 1997. The second system developed is called the Bond and Mortgage System that used the prototyping method of development, spanned mainly from the years 1996 to early 1998.

## The Development of the CAMFR System using the System Development Life Cycle

## (1990-1997)

In 1990 meetings were held with the CAMFR system managers and with the project manager and systems analyst to discuss the redesign of the 1990 CAMFR system. Management had already decided that the redesign of their system was the appropriate way to go for future automation and savings for the City of New York. Therefore, this researcher, the systems analyst on this project began with the Analysis Phase of the Systems Development Life Cycle in the development of the redesigned CAMFR system.

## The CAMFR System Front-End Redesign - Phase I (1990-1991)

The Contract Agency Monthly Financial Report (CAMFR) system provides the Office of Financial Management's (OFM) Division of Contract Agency Finance (DCAF) with a computer system to monitor funds disbursed to contract agencies and expended by them in the process of carrying out their contracts.

## Manual Workflow of the old CAMFR System (1990)

The objective of the old Contract Agency Monthly Financial Reporting System was limited to mechanizing the routine accounting and ledger functions associated with delegate agency programs, and providing a database for producing financial and management reports. Approximately 1000 Delegate Agencies were utilized to provide a wide range of Community and Social Services on behalf of HRA (Human Resources Administration), to both children and adults in 26 NYC poverty areas.

The monitoring and control of cash disbursement to delegate agencies for each of the five participating constituent agencies was historically done manually. OCAF maintained a manual ledger of cash expenditures against budget for all 1000 + delegate agencies. Some duplicate agencies also maintained a duplicate ledger. It was recognized that control of expenditures was not timely. The disbursement of funds was loose and too many times without proper documentation. Also, it was quite impossible to do cost analysis of expenditure against budget by HRA Fiscal across constituent agencies.

Attempts at redesigning the CAMFR system were attempted since the early 1970s but all these attempts failed. In the early 1970s a project was instituted to mechanize the CAMFR system. A very elaborate but unworkable and unmanageable system was developed. Early in 1977, it was recognized that the system developed did not serve the original purpose for many reasons including organizational responsibility changes. It was decided to retain the data entry system (4-Phase Mini) and redesign the system to serve the then current needs. In 1977, a project was undertaken by a Task Force at the request of HRA as part of their Management Plan to achieve the following benefits: reduction of clerical efforts, better management information, and improved internal control and accountability.

A computer system was developed using a mini computer and HRA's central processing computers to perform the following functions:

- Maintain agency master file

- Verify CAMFR submissions

- Prepare financial listings and management variance reports

- Prepare general ledgers

- Prepare next period CAMFRS


Each contract in the system created a master record. That record was modified every time the contract was changed and whenever the monthly expenditure data was entered.

On a line-by-line basis, reflecting the level of detail on the contract itself, the record contained the following information:

- Original Budgeted Amount

- Current Budgeted Amount

- Program Expenditures to Date

- Unpaid Bills

- Latest Projection

The CAMFR, or input form, was generated from the master record every time the latter was modified. It was, therefore, unique to each contract and accurately reflected the level of reporting and control desired by the constituent agency.

Whenever the CAMFR was produced, the system also printed a LEDGER. On a line (and cumulative) basis, the LEDGER displayed the budgeted amount, expenditures, unpaid bills, and projection. Those figures were compared to produce a series of computed fields. Certain computations (e.g. projections plus expenditures which were greater than budget) caused the item to be flagged. What criteria to use for flagging fields and even which fields to compute were decisions left to the constituent agencies.

The principal components of the old CAMFR system are summarized as follows: Files - The system contained two files, namely the Budget Master file and the transaction file. The heart of the CAMFR system was the Budget Master File, and is used as the source of all recurring and special reports. Separate records are kept for each Delegate Agency containing key descriptive data as well as a monthly and cumulative record of expenditure against Budget on an individual line item basis. The system also contained several related tables that further defined data appearing on the aforementioned files.

Input Data - Principal input to the system consisted of three documents: Monthly CAMFR Expenditure Reports submitted by the Delegate Agencies, including a statement of cash balances. CAMFR input form to register a new contract whenever a budget was approved to a Delegate Agency. And, an auxiliary input form to record the detail line budget amount pertaining to a new contract or a subsequent approved budget modification. Input data was entered via CRT devices to a mini computer located at 75 Worth Street, NYC.

Processing - The files were processed on a daily basis. Entry to these files was accomplished via CRT Unit processing which was the "Front End" of the system. The CRT edited and validated input data before it was placed on input file disks. Later the disks were spooled onto tape files for computer processing at 2 Broadway, NYC. The central processor maintained and updated all files and tables, as well as produced all reports of the system.

Output Reports - Output reports were produced monthly and on request by the user. These reports gave a detailed analysis of data processed including:

- Monthly CAMFR turnaround document which provided the agency with a blank form in which to provide monthly expenditure and projection information on a budget line item basis for each Delegate Agency.

- Monthly Ledger Report that displayed the budget amount, expenditure, unpaid bills, and projections on a cumulative line item basis for each Delegate Agency.

- Special Analysis and Variance Reports to highlight spending patterns against projections, as well as provide additional management information and evaluation reports.

Production - There were about 1,000 budgeted accounts and this database was updated on an ongoing basis during the month. Therefore about 50 budgeted accounts were affected daily over a twenty workday month. The Cathode Ray Tube (CRT) visual display units formatted records to conform to screen input documents. Once a record was completed or a screen was filled up, a physical record was written to disk storage that was a peripheral unit of the display processing system. The front-end mini computer accepted data and affected some data validation. The mainframe computer performed files updating, performed final data validation, and provided printed output reflecting Delegate Agency spending against budget.

## System Narrative of the Old CAMFR System (1990)

Data was input on a visual display device that formatted records to conform to the input documents. Check digit verification, field editing, and validation were features associated with the Data Entry System. The computing device used for data entry was a minicomputer. This system included a CPU with CRT (Cathode Ray Tube) keying stations. The CPU preserved all data entries on the disk while CRT operators keyed in data from the input forms. A daily transmission to the HRA computer center was effected so that the master file could be updated with current information.

Three computer-input forms provided the basic information source that was converted to machine-readable form via the data entry-keying stations. The first form permitted registration of a new delegate agency contract. The form allowed the unique ID#, agency name and address, plus the total contract (budget) dollars to be loaded in the CAMFR Master File. The second form was used to input detail budget line items (account number and dollars) to the CAMFR Master File. Both forms could be used to modify prior data on file. The third form, allowed an incorrect ID# to be corrected or deleted with or without deleting the associated data. The fourth basic form was the machine-produced "mechanical" CAMFR for reporting current monthly expenditures and subsequent dollar requirements.

The format of each computer input document was programmed as a CRT screen. When operators received the input forms, the appropriate screen then could be called up by the designated job name. Since the CRT screen was very similar to the input form, operators key in only the input data. A cursor under screen indicated the exact position of each entry. In addition, document totals, record counts, and error messages were used in data entry operations to ensure the accuracy and reliability of entered information.

Upon completion of document entry onto CRTs, source documents, batch transmittals and completed batch disposition summary were returned to data control for final reconciliation of batches transmitted via CRT with turnaround hard copy from 2 Broadway, to ensure that:

- All data transmitted had been processed

- Any discrepancies were reconciled

- All errors were resubmitted for correction

Upon making appropriate corrections, batch transmittals were prepared for processing. At the end of the processing cycle, CAMFR source documents were forwarded to Central File Room to be filed by contract identification number. The first and second forms were filed by administrative code and batch transmittals, and summaries were retained by Data Control in binder file in date order.

**Analysis of the old CAMFR System (1990)**

The CAMFR system in 1990 consisted of a front-end batch data collection system resident on a Motorola IV Phase minicomputer system located at DCAF's office at 200 Church Street in Manhattan, New York. This system consisted of 7 data entry programs, programmed in VISION, and produced transaction files written on 9 track tape that were used to update the master file.

The back-end of the system consisted of the master file and supporting tables, and was resident on the IBM mainframe located at Management Information System's (MIS) offices at 111 8th Avenue in Manhattan, New York. The master file was updated and reports produced whenever transaction tapes were received, usually on a daily basis.

## The Redesigned System in 1991

The front-end IV-Phase hardware was replaced by newer equipment, a Motorola super computer with a UNIX operating system, thus necessitating migration of the old front-end software onto the new equipment. To accomplish the migration, the VISION programs were analyzed and documented and revised programs were designed to duplicate the functions of the VISION programs and also provided the user with enhancements to both individual programs and the system in general.

Highlights of the Revised Phase I System were:

- Programmed in INFORMIX and SQL.

- Programmed as a menu-driven system.

- Use of a telecommunications line to transfer transaction files from the front-end to the back-end, thus eliminating the need for sending of tapes.

- Extensive on-line editing.

- Capacity for up to 10,000 transaction records in one transaction file.

- Ability of the system to accept all transaction records, then cull out those that cannot be sent to the back-end in the same transaction file, and send them in subsequent files.

- Downloading of a subset of one of the back-end tables so that edits can be performed and corrections made upon data entry rather than waiting for the master file update run on the back end for the editing to occur.

## System Design Specifications for the Redesigned CAMFR Front-End (1990)

- Data dictionary entries for each data element and elementary item; a total of 124 entries.

- An Appendix containing entries for each of the 16 files in the revised system, including the data fields contained, the record layout (where required in a particular order), and other relevant information.

- Programmer specifications with screen layouts for each program in the system, including data entry programs, table updates, and report production.

- Specifications for the telecommunications interface and data transfer between the front-end and back-end.

- Specifications for system maintenance and back-ups.

## Details of the Redesigned CAMFR System in 1991 - Phase I

## Menus

Menus are used to provide a user-friendly method of access to all system programs through a hierarchical selection process. This allows access only to authorized users of specific functions. The program contains six menus dividing CAMFR Front-end System functions into data entry, report generation, files transmission, table maintenance, and system maintenance. A main menu points to the lower level menus. System security is enforced through this program; users have to key in valid codes in order to access lower level menus and selections listed on menus.

## Data Entry

### NCAMFR

The NCAMFR program is used to enter basic identifying information about the Contract Agency and the contract. This program consists of three segments: NAME, ADDRESS, and CONTRACT. These segments can be run independently; the operator selects at the outset the segment(s) to be run. One, two, or all three segments can be selected. The Contract ID, batch number, and input selections are retained and displayed in all segments. Each segment of the program writes a record to the TRANS table.

The ADDRESS screen has a window to facilitate entry of address information. If the Contract Agency's city, state, and first three zip code digits appear in the window, this information is entered with a single keystroke. If this information is not in the table, it can be keyed into the appropriate fields. After data entry on a given Contract ID is complete, the BUDGET program can be accessed from NCAMFR's ring menu. If BUDGET is selected, the BATCHNO and CONTRID values are passed to it.

### BUDGET

The BUDGET program is used to enter expenditure amounts allotted to specific categories of expenses within a given contract. Initial budgetary allotments are input at the start of the contract; revised allotments are input as needed throughout the life of the contract.

Data collected by this program consists of budget line codes that identify an expenditure category, the expenditure amount allotted to each budget line, and an associated reimbursement claim category. Data entered through one pass of this program can result in multiple records being written, as each unique budget line code and its associated fields result in one record. A variable number of budget lines, from 1-150 can be entered during one program pass. At the beginning of the program, the operator keys the number of budget lines that will be input into RECNO. This number is used by the program to determine the number of budget lines to display on the screen and in the formula for computing how many records that would ultimately be written.

In addition to multiple budget lines, there is also a 'Total' line. The value in BUDLN2 for the Total line is computed by the program from the sum of all budget line amounts and displayed on the screen. The 'Total' line fields are also written into a record. Line numbers are displayed on the screen for the operator's benefit; they are not written to the record. Fourteen budget lines can fit on one screen. Whenever more than 14 budget lines are to be keyed in, it will be necessary to scroll this portion of the screen to accommodate all the lines. The screen can also scroll backwards to enable corrections to previously entered lines.

Upon input, budget line codes are edited in two ways: 1) Against BUDTBL, the table containing valid values for budget line codes to assure that a valid value is input, and 2) Against previously put budget line codes to assure that a duplicate code is not input. Records written by this program are written in the BUDGET segment of the TRANS table.

<u>CAMFR</u>

The CAMFR program is used to record a Contract Agency's monthly expenditure information and to reconcile the receipt and disbursement of funds received from HRA. This program consists of two parts, corresponding to the two principal sections of the CAMFR document. The first section is used to enter detailed expense information pertaining to the Contract Agency. The second section reconciles receipts and disbursements. The expense categories in the Expenditure section vary from agency to agency; the categories in the Reconciliation Section are constant. To facilitate data entry, budget codes used in the Reconciliation Section are hard-coded. To avoid entering partial information, the program will not write expenditure records without reconciliation records or vice versa.

<u>B-C-D</u>

This data entry program is used to create BLANK CAMFR, CASVR, and DELETE CAMFR record types through the use of one program. A menu format provides for selection of the desired action: to create a blank Camfr, delete a Camfr, or create a CASVR. One screen is used to capture data for both the blank Camfr and deleting of a Camfr. A different screen is used for capturing CASVR data. Only authorized personnel can enter the 'Delete' module. This is enforced through a SECURITY check when this function is selected, with entry to the module denied if unauthorized. Records can be keyed in and written into groups of 12 at a time.

## SECURUPD

This data entry program is used to add, modify, delete, and query records in the SECURITY table. Descriptions of the four different functions of the SECURUPD program are the following: ADD - New records are added. UPDATE - Record fields can be modified and rewritten. REMOVE - Records can be deleted and the SECURITY file repacked. QUERY - Records can be viewed sequentially or by USERID.

Access to components of the CAMFR Front-end is controlled positionally via the PGMS field, with a 'Y' indicating those components acceptable to the user. When total access is granted by keying in 'Y' in PGMS byte position #1, the other 6 bytes are program filled. If total access is not granted the cursor moves sequentially from bytes 2 through 7 for key-in of desired accesses.

## ADMCDUPD

The purpose of the ADMCDUPD data entry program is to maintain records in the ADMINTBL. This program provides for adding new records, deleting existing records, listing the contents of the table on the screen, and printing it out. Selection of the specific action desired is made at the Ring menu prior to entering data.

## ZPCDUPD

The purpose of the ZIPCDUPD data entry program is to maintain records in the ZIPTBL. The data entry program provides for adding new records, deleting records, listing the table on the screen, and printing it. The list and printout show CITY, STATE, ZIP3 once and as many ZIP2s as there are belonging to that CITY and ZIP3. ZIPTBL records are sorted by ZIP3, ZIP2, STATE, then CITY.


## PGMACTUPD & JE5BUDUPD

The purposes of the PGMACTUPD and JE5BUDTBL data entry programs are to maintain records in the PGMACTBL and JE5BUDTBL respectively. These programs provide for adding new records, deleting existing records, listing the contents of the tables on the screen, and printing them. Selection of the specific action desired is made at the Ring Menu prior to entering data.


## **Mainframe CAMFR System Revisions - Phase I (1991)**

The purpose of the specifications written for this part of the SDLC project was to provide logic and guidelines for revisions to the back end (IBM) CAMFR system that enables it to interface with the revised CAMFR front end.

The redesigned CAMFR front-end system does not send transaction files to the mainframe via 9 track magnetic tape anymore. Files are transmitted via a telecommunications line connecting the Unix hardware at 220 Church Street to the IBM mainframe at 111 8th Avenue. CAMFR transaction files (used as inputs to mainframe CAMFR system processing) are transmitted directly onto DASD at 111 8th Avenue. Changes required in mainframe CAMFR system production processing are a result of the revised storage media. The major areas affected by this change were:

Two permanent disk files are created as production data sets on the IBM system for storage of CAMFR transaction files: PROD.OCAF - This file is used for storage of transaction files used in the CAMFR system master file update and deletion runs, and blank CAMFR production. PROD.CASV - This file is used for storage of CASVR1 files. Under the old CAMFR front-end system, CASVR1 files are sent on tape with the DSN = HFSCASVR, and the VOLSER = OCAF01 through OCAF20. Under the revised system, CASVR1 files are stored on the disk under the above DSN. The disk file capacity is 3,000 records.

In addition to the two new production files, two temporary disk files are needed for testing purposes, one file for OCAF, the other for CASV. Once transmitted and stored on the mainframe, the OCAF and CASV files could be backed up onto 9-track tape. Backup versions are available if a rerun is needed.

## Mainframe Run Scheduling

OCAF - The master file update and related processing are scheduled for daily execution as a production run. Therefore, it is no longer necessary for a tape to be mounted before execution, the disk file is accessed instead. The user uploads a new file late in the day or early evening, therefore, mainframe scheduling of the production run should indicate night time or early morning execution.

CASVR1 - The CASVR1 File continues to be sent on an ad hoc basis, only a few times a year. Thus this run continues to be performed on an as-needed basis. When a CASVR1 file is uploaded to the mainframe, the user sends in a memo requesting processing, just as they did in the past. The only difference is that the input file is now accessed from disk, rather than tape. Each subsequent CASVR1 file uploaded to the mainframe overwrites the previous file.

## **Highlights of CAMFR System Redesign - Phase II (1992-1993)**

In order to automate the processing between the master file at the back-end on the mainframe computer and the transaction file at the front end with the UNIX LAN, the following specifications were written and implemented by the programmers. This was done to automate data transmission from lower Manhattan at the front end to the back end mainframe computer at 111 8th Avenue.

EXTCONVT: Master File Extract Conversion Program (UNIX to Informix) Timer Version.

The purpose of the EXTCONVT program is to convert the three Master File Extract files (Contract. Unx, Financial. Unx, and History. Unx) that are downloaded nightly by the IBM mainframe at 111 8th Avenue to the Motorola Unix system at 220 Church Street and stored as Unix databases into Informix tables.

Two versions of this program were developed:

1. EXCONVT - Containing a timer so as to run automatically.

2. EXTCONVM - Menu version without timer. This version can be invoked through a menu selection (for times when the timer version doesn't work).

In order to have the files updated and ready for the user each morning, this run has been put on a timer. At 7:00 A.M. each morning the system clock triggers the run. Each of the three files is read in turn. If there are records to be converted in the Unix file, the existing Informix version is deleted and then recreated and the Unix version is converted into the Informix table. A record is written to the EXTRACTLOG file indicating the success of the conversion and the Unix version of the file is then deleted. If there are no records to be converted (indicating that downloading was not successful or not attempted) a message is written to the EXTRACTLOG indicating that the Unix file was not found.

**Highlights of the CAMFR System Development Project - Phase II (1992 - 1993)**

HISTORY-PGM

Analysis was done, specifications written, and programming code completed for the HISTORY-PGM which is used to create an extract file of summary data on closed CAMFR contracts stored in the production CAMFR master file. The file is created on the IBM mainframe on an ad hoc basis and then downloaded onto the UNIX system at 220 Church Street, the front end of the CAMFR system.

CONTRACT-FINAN-PGM

Analysis was done, specifications written, and programming code completed for the CONTRACT-FINAN-PGM which is used to create two extract files, the MFEXTRACT.CONTRACT and the MFEXTRACT.FINAN file, from data stored in the production CAMFR master file. These files are crated on the IBM mainframe daily as part of the CAMFR master file production run, and are downloaded to OFO's UNIX System at the front end of the CAMFR system.

EXTINQ

Analysis was done, specifications written, and programming code completed for the EXTINQ program. The purpose of the EXTINQ program is to provide inquiry access on the UNIX front-end system to CONTRACT, FINANCIAL, and HISTORY file records extracted from the CAMFR master file on the mainframe. This program provides display screens as a means of accessing CONTRACT, FINANCIAL, and HISTORY records; a separate screen is used for each type of record. Depending on the type of record, the user accesses the data by entering Contract ID, start date, end date, delegate name or IRS code.

The program selects and displays the appropriate records on the screen. Data are displayed from CONTRACT, FINANCIAL, HISTORY, and ZIPTBL tables.

For CONTRACT and HISTORY records, groups of records can be selected and browsed by specifying appropriate selection criteria. A Browse menu allows the user to navigate through the records selected. A Help screen for the Browse option provides additional information.

CUMEX

Analysis was done, specifications written, and programming code completed for the CUMEX program. This program crates a report showing the cumulative expenditure for each type of expense within a given administrative code. The report appears in two sections. The first contains data for each admin code in admin code order. The second groups the admin code by program: CDA, ACD, CWA, etc.

The following elements from the FINANCIAL INFORMIX table are used in this program: (1) CONTRID, the Contract ID. The first three bytes of CONTRID, the program year and administrative code, referred to as PGMYR-ADMINCODE, are also utilized; (2) BUDLN, which indicates the type of expenditure; And (3) EXPNDAMT, the cumulative expenditure for a particular BUDLN.

## **Highlights of the Development of the CAMFR System - Phase III (1994)**

The purpose of CAMFR Phase III was to revise the entire CAMFR system in order to accommodate larger values in money fields. This was accomplished by enlarging the money fields by removing the decimal point, thereby converting the existing cent bytes to dollar bytes. A general description of the analysis, specifications, and programming code completed for this phase follows:

CAMFR Back-End

1) All money fields were identified in all CAMFR programs.

2) All money fields were identified in files.

3) Specifications were written to convert data in files.

4) Specifications were written to convert programs.

CAMFR Front-End

(Data Entry and Master File Extract)

1) All money fields were identified in all programs.

2) All money fields were identified in all files.

3) Specifications were written to convert programs.

4) Specifications were written to convert files.

5) Specifications were written for conversion of data files, if necessary.

The entire CAMFR system, both the back-end and front-end, programs, data files, and data fields were changed simultaneously. This effort was synchronized and coordinated across both IBM and UNIX programmers.

## **Highlights of the Development of the CAMFR System Phase IV (1995 - 1997)**

Analyses, design, specifications writing, program coding, and implementation of CAMFR Phase IV concentrated on connecting data from the front-end and back-end of the

CAMFR system and the Year 2000 coding changes. The following programs were revised or created during CAMFR Phase IV:

NCAMFR

The purpose of the revised NCAMFR program is to connect data in the Contract Extract file with the NCAMFR data entry program in order to reduce key-in and improve integrity of data in transaction records created on the front-end. It is used to enter basic identifying information about the Contract Agency and the contract. Year 2000 logic has been included for the contract screen area.

This specification builds on old specifications, the logic holds except where additions were made. The data entry screen has been revised. NEW is to be selected when a data entry of a NEW Camfr or a RENEWAL Camfr has to be done at the beginning of the Contract Year. MOD is to be selected when a modification to the existing current year's NCAMFR has to be done during the year.

The BATCHNO is to be retained as long as the user is in the NCAMFR program, and data entry continues. It can be modified or retained on key in of a new CONTRID. With a NEW-RENEWAL data entry and a MOD - the most recent Contract records for the keyed in Contract ID are displayed on the data entry screen for acceptance modification of Je5Codes and amount. With a contract that is completely new to the system, the entire data entry including data entry of JE5 Codes and amount must be done.

After data entry on a given Contract ID is complete, the BUDGET program can be accessed from NCAMFR's ring menu. If BUDGET is selected, the BATCHNO, CONTRID, and JE5CLAIM1 values are passed to it. In the Contact area of the screen FYR and LYR have both been changed from 2 bytes to 4 bytes to accommodate the century in the year and for 'less than' or 'greater than' comparison purposes.

BUDGET

The purpose of the redesigned Budget data entry program is to connect data in the Contract and Financial Extract files with the Budget data entry program to reduce key-in and improve integrity of data in transaction records created on the front-end.

At the Ring Menu of the Budget Data Entry Screen are NEW, BUDMOD, and EXIT. EXIT is to be selected when the data entry operator wants to exit from the Budget data entry screen. NEW is to be selected when a data entry of a NEW Contract Budget or a RENEWAL contract Budget has to be done at the beginning of the Contract Year. BUDMOD is to be selected when a modification to the existing current year's Budget has to be done during the year.

The BATCHNO should be retained as long as the user is in the BUDGET program, and data entry continues. Programming has allowed for it to be modified or retained on key-in of a new CONTRID. The system would know when it is a RENEWAL Budget because the Contract ID will be in the system within the last two years.

With a NEW-RENEWAL data entry and a BUDMOD, the most recent Financial records for the keyed in Contract ID are displayed on the data entry screen for modification of Budget Codes, amounts, or JE5CLAIM2. With a contract that is completely new to the system, data entry of budget line codes and amounts must be done. Some fields may be passed from the New CAMFR Contract data entry program or CONTRACT Extract table, such as the JE5CLAIM1 and BATCHNO.

The RECNO is automatically calculated at the end of data entry, and the user has to compare it to the input RECNO value. If a problem, modifications are allowed on screen. The total dollar amount is automatically calculated and if correct, records are written to the Budget file, if incorrect user can correct screen values.

## Y2K Coding Fixes

For the BUDGET program some changes had to be made for the calculation of the years that are to be the current years and the years that are to be future years. The CAMFR system is coded according to fiscal years beginning every July 1. Some of the years in the CAMFR programs are only 4 digits for MMYY. So, in 1996, the programmers and systems analysts on this CAMFR system realized that there would be a problem of going from 1999 to 2000, and from 2000 back to 1999 if the use of two bytes for YY was to continue. Calculations from 1990 to 1999 were as follows to arrive at the next fiscal year: $90+1 = 91...98+1 = 99$. The question was how to get from 99 to 00 and 00 to 01 when there was a '19' hard coded in the programs. So in reality the results were incorrectly calculating as follows: $99+1 = 100$.

This researcher as a systems analyst on the project suggested the following to be done to solve this Y2K date problem and wrote specifications for the programmers to follow in order to re-code the programs. If the two-digit year 99 is encountered in a program, it will be automatically changed to -01, therefore, the fiscal year calculation for the next year would be $-01+1 = 00$, and the two bytes would remain for the year 00. The programmers then coded all programs with two bytes for the year using -01 for 99 in every instance.

Another reason why the two-byte year 99 was logically replaced in the programming code with the two-byte year -01 was because of the calculation of the current contract year. If a contract year began in 98 and ends in 99, then the subtraction of the start-year from the end-year would tell whether the contract year is still in effect or if it has expired. Therefore, by doing the calculation:

$$99 - 98 = 1$$

would tell the program and the users that the result of '1' would say that the contract is current. This calculation is a problem if the start-year is 99 and the end-year is 00. The calculation would have then been as follows:

$$00 - 99 = -99$$

which is the wrong answer when '1' was needed to determine if the current contract year was still in effect. Therefore 99 was replaced in the program code with -01 and the following calculation could then be done:

$$00 - (-01) = 1$$

and the correct answer of '1' is obtained to tell us that the current contract year was still in effect.

Therefore, the COBOL programs doing this calculation of current year did not have to change the number of bytes of year from two bytes to four bytes. Two bytes were kept and the logical changing of 99 to -01 was done everywhere in the data when the two-byte year 99 was encountered.

However, where there was a four-byte coding for year the programs were changed to be able to not use a hard code of 19 as was done throughout the 20th century but the bytes were allowed to accumulate to 2000 or even the year 3000 or above. This was done as a precaution, just in case the programs were still around and up and running successfully in the year 3000 or later. This would then take care of the year 10,000 or even 100,000.

NEWCAMFRX&R

The purpose of the NEWCAMFRX&R program was to connect data in the Contract and Financial extract files with the monthly Camfr data entry program to reduce key-in and improve integrity of data in transaction records created on the front-end.

This is an enhancement of the old CAMFRX and CAMFRR CAMFR data entry program. Enhancements included are selections on the Ring Menu for Single Camfr or Multiple Camfrs to be keyed in. This program allows for less fields to be keyed in. Fields such as the Report Date, Budget Line Codes, and Opening Balance are derived from the Contract and Financial extract files at the front-end. Another enhancement over the old CAMFR data entry program is that while data entry is not yet completed, any money field could be corrected before the CAMFRX or CAMFRR transaction records are written to the TRANS file.

The logic of data entry could be simple or complex. With complex logic, the Report Date is keyed over, thus not following directly to the Contract extract record date of last Camfr, in this case the Opening Balance is also keyed over. In the simple logic, the derived Report Date displayed is used and the Opening Balance does not have to be keyed over.

At every phase of the CAMFR system project implementation, the users were trained and user manuals were developed. See Appendix A for figures of CAMFR system screens and excerpts from a user manual. In 1990 there were approximately 30 data entry operators at the front-end for processing the CAMFR system. At the beginning of 1997 at the final implementation of the redesigned CAMFR system there were approximately 3 users at the front-end and one programmer at the back-end who used the automated system to run the CAMFR system and thus took the place of over 30 workers. Most of the data entry and other processing were automated therefore there was less need for data entry and physical workers.

## The Development of the Bond and Mortgage System using the Prototyping Method of System Development (1996-1998)

In 1996 meetings were held with the Bond and Mortgage System managers, users, the project manager, and the systems analyst, this researcher, for the development of a prototype for the automation of the manual processing done by the Bond and Mortgage Unit. Automation of the Bond and Mortgage Unit was requested so that the City of New York could realize more monetary receipts. Therefore, the feasibility part of the system development was already done and the managers and users agreed on having the new automated Bond and Mortgage

System developed in the form of a prototype at first.

## Bond and Mortgage System - The Existing System (1996)

The Office of Revenue and Investigation, Real Property and Assets, 60 Hudson Street, 8th floor, New York, NY 10013 in the Human Resources Administration of the City of New York is responsible for the assignment of monies from the sale of real property owned by Public Assistance and Medicaid clients.

The current staffing is 6 people including 2 temporary workers. There is a title examiner, a Public Assistance worker, a Medical Assistance worker, a typist, a receptionist and a personal computer worker. Interviews are done on Tuesdays and Thursdays. During the other three days general work is done on the cases. In 1996 the Bond and Mortgage Unit collected about $1 million per year.

All work was done manually; letters were xeroxed and names and addresses and other information were handwritten. However, some letters have been put on the personal computer and the details typed in. The volume of new work per week averaged about 30 cases. There was a Unix computer, one personal computer, and a word processing typewriter in the unit. However one staff member had access to a second personal computer and printer.

A printout existed of part of the old database that was destroyed, showing the codes that were used, the case name, case number and other pertinent data used by the Bond and Mortgage division. The personal computer was mainly used to produce reports. Daily reports were produced on title searches, Medicaid cases, and a list as to what went over to the litigation unit.

## Bond and Mortgage Automated System Needs (1996)

The user wanted to automate the system from the point when the intake system worker referred the case for an appointment or a Medical Assistance Payment referral on an institutionalized individual was received. The automation was to end when the payment was received and the satisfaction document was given for the payment of the lien.

The user also wanted the following:

- A database

- Calendar Scheduling

- Printing of Bond and Mortgage Forms - Legal Forms and Letters

- Matches to other Systems - Finance and the Welfare Management System etc.

- Connection to the Collection System

The user also wanted the ability to pull up cases in the following ways for houses and condos:

1. Alphabetically

2. Numerically

3. Borough

4. Lot #

5. Block #

In addition, the user wanted to pull up cases similarly for co-ops, and to inquire on data. Basically the user wanted to have more help processing the workload. Also, the user needed flexibility built into the system for their future needs such as: to produce matches, reports, and other inquiries.

## Overview of the Bond and Mortgage Process (1996)

1. New appointments came in from the Public Assistance centers. The receptionist made the appointments. A minimum of 10 to 15 appointments is processed twice a week on Tuesdays and Thursdays. The supervisor explained to the people what it involved.

2. The clients signed necessary papers and then went to the title examiner. The title examiner set up all the papers for the typist to type and to have a copy recorded.

3. Everything was proofread. There was a service that took the title searches out to all boroughs in New York City to the local register's office, except the Bronx.

4. There was a wait time for them to come back. The Brooklyn county clerk's office was 1 1/2 years behind.

5.  Once the recorded copy is returned, cards were completed, three cards for property and two cards for co-op.

6.  If someone wanted to sell a house or refinance a house they had to request that we give a letter. They would come in and ask us for a payoff letter. Often it was the lawyer, Title Company, or the client.

7.  A Welfare Management System print out from 111 8th Avenue was then obtained.

8.  A letter was sent out to whomever sent in the inquiry as to the exact amount that was due.

9.  A request to subordinate or negotiate a payoff figure may be the response to the payoff letter.

10. It usually took 30-60 days for the bank to give a commitment.

11. If a response was received, it usually was a check. Checks were sent to the lock box.

12. The check was put through as per the disposition form.

13. When the check cleared, if it was not sent to the lock box, we made up a satisfaction of mortgage, lien, or chattel mortgage in the case of a co-op.

14. A response would then come back from the Title Company and a satisfaction was recorded.

15. The case was then filed in the closed satisfied file.


## Queries Currently Done (1996)

1.  In Rem is a city foreclosure for nonpayment of 3 years of real estate taxes. By borough, it was brought up on the Department of Finance computer terminal screen from lowest block number on the borough.

2. Breakdown by borough, hunt for lot and block, go by name or lot and block if there was a name change with it.

3. There was a Department of Finance access from computer terminals on the floor of user's office that showed current owner only.

4. Inquiries regarding liens were made on the Department of Finance computer terminal. Who put the lien on was not available, by date we assumed it was our lien.

5. The title searcher went out to the County Register's Office and got not only a complete printout but went back as to who owned it from the date it was bought.

6. One title search per case was done up to the date that the Bond and Mortgage Department was doing it.

7. In Rem agreement could be applied for and the title regained by the client.

## Bond and Mortgage Proposed Overall System Design (1996)

Appointment scheduling was to be done by the Microsoft Schedule package and then linked to the Schedule File where records were to be written and stored. Microsoft Access was to be the data entry and database format to be used. Data was to be entered through Access data entry screens. Microsoft Word was the word processor to be used to fill in forms and letters, merging through the Microsoft Word database from the Access database files fields. Design and generic specifications of individual data entry programs were to follow; the first to be worked on was scheduling.

## Bond and Mortgage Prototype System Design (1997)

Using Microsoft Office 95 Professional, then Microsoft Office 97 tools and capabilities, a prototype of the Bond and Mortgage System was developed on a floppy disk. This prototype was developed by the systems analyst, this researcher, at a time when the user did not have any Pentium computers at the work site. One Pentium computer was then installed at the user site and the prototype was then demonstrated there. Since more Pentium computers were planned for the user site the Bond and Mortgage System prototype was developed with this in mind to take the place of the manual processing that was taking place in 1996 and 1997.

The Bond and Mortgage System prototype floppy disk contained the following four main folders: Bond and Mortgage Appointments, Bond and Mortgage Excel Files, Bond and Mortgage Word Documents, and Bond and Mortgage Access Files. Test data and programs in these folders were all coordinated and intermingled to work with each other.

The following skills in working with Microsoft Office 97 and the Pentium personal computer with the different parts of the Bond and Mortgage system were taught to the user and a user manual developed for their use:

- Putting the floppy disk folder onto their computer containing the system prototype so they could practice with it before the system was approved and implemented. This was done so that there was feedback as to what was being developed in order to obtain suggestions for further improvement of the Bond and Mortgage System.

- Using the Bond and Mortgage floppy disk, and also being able to run the floppy disk directly from Drive A.

- Loading the Bond and Mortgage floppy disk onto the computer desktop and using it from the Bond and Mortgage folder to do test cases.

- How to work with the Appointments Folder which at that time was in Microsoft Schedule +.

- How to work with Microsoft Access to create and maintain a database of all the Bond and Mortgage clients.

- How to merge Word documents with records from the Access database in order to create letters to the clients.

- How to work with Microsoft Excel to do calculations and spreadsheets for the Bond and Mortgage System.

- How to add fields to the Bond and Mortgage master file which was created in addition to a data dictionary and different files for the different processes of the Bond and Mortgage System.

- How to delete fields from the Bond and Mortgage master file.

- How to create files other than the master file or history file.

- How to create a data entry form based on a query file in Microsoft Access in order to enter data about clients into the Bond and Mortgage database.

- How to create an appointment file folder for the Bond and Mortgage System folder.

- How to enter data into query forms or data entry forms to be written to Access database files.

- How to create a new correspondence history table for duplicate case numbers.

- How to create merged Word documents so that any letter or document produced by the Bond and Mortgage system could be merged with data from the Bond and Mortgage Access database in order to send forms and letters to the clients. See Appendix B for a sample of how a Bond and Mortgage form letter was merged with data from the Bond and Mortgage Access database.

- How to actually merge database records from a Microsoft Access Bond and Mortgage query using a Word document.

- How to create reports from Access files or queries.

- How to create a correspondence history report

- How to create Microsoft Excel spreadsheets from Access files or queries.

- How to set up a security system in Word documents.

- How to set up security in Access database files.

- How to calculate totals of recurring fields from files or forms on Excel worksheets.

- How to make a copy of a folder on screen.

- How to copy a folder to a diskette.

- How to delete a field from a query or file.

- How to insert a field from a query or file.

- How to delete a field from a table.

- How to add a field to a table.


## Bond and Mortgage System Implementation (1998)

The user approved the final prototype diskette of the Bond and Mortgage System in early 1998 and then it was uploaded to a network folder where all users of the Bond and Mortgage system could access it from their Pentium computers. This system then grew into a larger interactive online system which was used by the user to add records to the Bond and Mortgage Access database and to create all documents needed by the system in Microsoft Word for merging with records in the Bond and Mortgage Access database. This automated system was able to keep and process records for all past, present, and future clients of the Bond and Mortgage System. This could not be done before. As a result the Bond and Mortgage Unit was able to recoup millions of dollars in liens from clients who were falling through the cracks with the manual system. Monetary receipts and profits of the new automated system thus increased by over 100%.

## Data Analysis

### Data Analysis for the CAMFR System Development

All throughout the development of the different phases of the CAMFR system meetings were held with user management, users, the project manager, and the systems analyst, this researcher, to ensure that the project was on the right track. For all implementation phases user signoff was received from user management to confirm that the system worked successfully.

The users of the CAMFR system were the City of New York, Human Resources Administration, employees who worked in the Department of Contract and Accounting Finance (DCAF). The user managers also worked in DCAF and oversaw the whole CAMFR daily operations. The project manager and systems analyst of the CAMFR project were also The City of New York, Human Resources Administration employees, who worked in the Management Information Systems Department. The programmers were also HRA employees who worked in the Technical Services part of HRA-MIS.

The fact that everyone involved in the CAMFR system development project was an HRA employee was noteworthy because the existing CAMFR system of 1990 was developed by consultants who had been paid on contract by the City of New York, Human Resources Administration. The redesigned CAMFR system, designed and implemented from 1990 through 1997, was developed by HRA MIS personnel who did not have to be paid extra for their time - just their annual salary unless overtime came into play. This resulted in The City of New York saving more money because The City did not have to pay extra consultant salaries.

Because only HRA employees were involved in the development of the CAMFR system, there was a sense of trust and an eagerness to impart knowledge by the MIS employees and the openness for the receipt of knowledge by the users. The users ended up learning the system so thoroughly when it was implemented that the CAMFR system now runs successfully everyday without any input by the MIS computer employees.

The CAMFR system development was slated to take several years because user management believed that past attempts to automate the CAMFR daily processing failed to encompass all areas needing to be automated. This had resulted in an incomplete and somewhat unsuccessful system being built which was unprofitable for the City of New York. Therefore, the Systems Development Life Cycle method was chosen for this project.

The Front-End of the CAMFR system was totally redesigned. The front-end encompassed the DCAF unit and the data entry operators using the LAN system there. The Back-End of the CAMFR system was redesigned somewhat to enable overnight processing of transaction data from the front-end. The back-end encompassed the MIS department and the IBM mainframe computer used to process the daily work from the CAMFR system.

One notable redesign of the CAMFR system was that instead of using trucks to deliver tapes of transaction data produced by the front-end, telecommunications lines were used to transmit the data instantly. Everyday based on the time most of the data-entry work was completed the upload took place from the front-end to the back-end. Another noteworthy enhancement that took place was that reports did not have to be printed and delivered from the back-end to the front-end using trucks every day. The reports were automatically printed at the front-end now using telecommunications lines from the back-end mainframe to the front-end printers.

The Y2K problem was going to stump the processing of the current contracts and past contracts of the CAMFR system if something was not done to take care of the Y2K coding changes, especially the COBOL programs coding changes on the back-end IBM mainframe. The systems analyst, this researcher, analyzed the problem that was going to be encountered in the year change from 1999 to 2000 and came up with the Y2K coding changes in 1996 described above. This prevented the CAMFR system from incorrectly processing current, future, and past contracts in the years 1999, 2000, and 2001. The coding changes would last for centuries even past the year 100,000 or 1,000,000. This was done to ensure that the programs would work successfully no matter how long they are in use, even if it is for years or centuries.

At each CAMFR implementation phase the users were trained by the systems analyst, this researcher, and the new software installed. At the end of the first phase new hardware was installed which were more modern and processed the work faster and were user-friendlier than the older computers. At the end of each implementation phase this researcher and another professional technical writer created user manuals for the users to refer to when doing their daily work on the CAMFR system. A more detailed supervisor's manual was also developed for the supervisors and managers of the CAMFR system. Once or twice the users telephoned MIS to ask questions about processing their work using the CAMFR system but after a short while they were comfortable with the system and able to use it without any outside help.

## Data Analysis for the Bond and Mortgage System Development

All throughout the development of the prototype for the Bond and Mortgage system meetings were held with user management, users, the project manager, and the systems analyst, this researcher, to ensure that the project was on the right track. For all versions of the Bond and Mortgage prototype floppy disk user approval was received from users and user management to confirm that the system was progressing correctly.

The users of the Bond and Mortgage system were the City of New York, Human Resources Administration, employees who worked in the Office of Revenue and Investigation (ORI). The user managers also worked in ORI and oversaw the whole Bond and Mortgage daily operations. The project manager and systems analyst of the Bond and Mortgage project were also The City of New York, Human Resources Administration employees, who worked in the Management Information Systems Department. The programmer of the system was the systems analyst, this researcher, who used Microsoft Office Professional 1995 and then Microsoft Office Professional 1997 to develop the prototype.

The fact that everyone involved in the Bond and Mortgage system development project was an HRA employee was noteworthy. This was so because the Bond and Mortgage system, designed and implemented from 1996 through 1998, was developed by HRA MIS personnel who did not have to be paid extra for their time - just their annual salary unless overtime came into play. This resulted in The City of New York saving more money because The City did not have to pay extra consultant salaries. However, at the implementation of the system in 1998 an outside consultant was hired for the maintenance of the Bond and Mortgage system.

Because only HRA employees were involved in the development of the Bond and Mortgage system, there was a sense of trust and an eagerness to impart knowledge by the MIS employees and the openness for the receipt of knowledge by the users. The users ended up learning the system so thoroughly when it was implemented that the Bond and Mortgage system now runs successfully everyday without any input by the MIS computer employees.

The Bond and Mortgage system development was slated to take only one or two years because user management believed that because of the development of office products like Microsoft Office Professional, the work of the unit could easily be automated with Microsoft Office tools and capabilities. Therefore, the user and user management requested a customized use of the Microsoft Office software for the Bond and Mortgage System. This was at a time when Microsoft Corporation and its many productivity and other systems software were becoming very popular. Everyone wanted to use Microsoft's software products that were renowned to be the best for office workers. Therefore, the prototype development method was chosen for this project. Prototyping was used also because the users did not have a Pentium computer to use at the beginning of the Bond and Mortgage systems development project. This researcher (the systems analyst) used a laptop to demonstrate the prototype at the user site. At the later stages of system development then the users began obtaining Pentium computers for use at work and they were able to test out the Bond and Mortgage prototype at their own computers. This occurrence sold the project and prototype to the users more quickly.

The entire manual processing of the Bond and Mortgage system was placed on the Pentium computer using a floppy disk prototype which was expanded later into a network client/server system. The users accessed the Bond and Mortgage folder from the network and are able to process all their work in the network folder that contained the various parts of the Bond and Mortgage system. Today the system is much larger than the size of a floppy disk.

One notable design of the Bond and Mortgage system was that the work was no longer performed manually. Every aspect of the Bond and Mortgage system was placed on the computer. This allowed for better record keeping, more efficient processing of daily work, and faster and more receipt of monies due from lien clients to the City of New York. Even before the final implementation of the prototype of the Bond and Mortgage system the Bond and Mortgage Unit was receiving a profit. Over 100% increase in the receipt of funds from clients was realized while the prototype was being developed and the users were beginning to use various ideas and parts of the prototype in their daily work.

At each major addition to the Bond and Mortgage prototype the users were trained by the systems analyst, this researcher, and the new prototype software installed. At the end of the first phase new hardware was installed which were more modern and processed the work faster and were user-friendlier than the older computers. At the end of each prototype development phase this researcher (the project leader) created user manuals for the users to refer to when doing their daily work on the Bond and Mortgage system. A more detailed supervisor's manual was also developed for the supervisors and managers of the Bond and Mortgage system. Once or twice the users telephoned MIS to ask questions about processing their work using the Bond and Mortgage system but after a short while they were comfortable with the system and able to use it without any outside help.

## Summary

This chapter showed the analysis, design, and implementation of the two systems developed and implemented successfully at New York City's Human Resources Administration using two different approaches to systems development. The two methods of systems development were the Systems Development Life Cycle and Prototyping. The next chapter will summarize and discuss the study and will also present recommendations for future research.

*Dr. Jennifer W. Gilmore, PhD*

# SUMMARY, DISCUSSION, AND RECOMMENDATIONS

## Chapter 5

### <u>Summary</u>

The main results of this case study are it showed how the two systems development projects (CAMFR and Bond and Mortgage) were developed and implemented successfully and are still in use today. The different stages of the Systems Development Life Cycle were used in the CAMFR system to develop and implement the CAMFR system successfully. The stages went from feasibility which was already completed when the MIS staff took over the project to analysis, design, developing of the code, testing, implementation, user training, development of user manuals, and maintenance of the system. The different stages of Prototyping development were used in the Bond and Mortgage system to develop and implement the Bond and Mortgage system successfully.

Each step in development of these two systems was done with the approval of user management, users, and the computer systems personnel at MIS. Because manual workflows were completed at the first stage of each project it was easier for the MIS workers to understand what the units were doing and were able to put the processing of the work on computer for more efficient job processing by the users of both systems. Also, because management and users along with the MIS workers were all involved at the beginning, this ensured that the projects would be accepted and successfully implemented.

Interviews with the current users of the two implemented systems reveal that the systems are still running successfully and that it has improved the work of the units by over 100%. The CAMFR system was also cloned to work for another agency, the Agency for Children's Services (ACS). ACS is using the CAMFR system to process their financial contract work just as HRA is using the CAMFR system to process their financial contract work. The Bond and Mortgage unit is now using the Bond and Mortgage system to process every aspect of their work. The computer and the skills learned by the users while training for the automated Bond and Mortgage system are being used in many aspects of everyday life even at work and at home on home computers, many users reported.

In both of these project development, the MIS personnel, especially the systems analyst, this researcher, had to physically go to the user site and train the users to use the computer and the new data-entry screens. At first, the users had to be trained in the use of the new computer hardware and then they were trained in the specific applications developed on the computer to process their daily work. Since these computers were ordinary computers, other things could be done with them including word processing for other jobs that the users may have and also calculations and e-mail communication with other employees at work.

This study supports previous research on how systems development projects proceed and are successfully implemented with the co-operation of users, management, and MIS systems development workers.

## **The Answer to the Research Question**

The Research Question

Are there more advantages than disadvantages in the development of business software computer systems using the systems development life cycle method and the prototype development method or are there other methods of systems development that are better suited to business environments?

The Answer

The advantages of the two methods of systems development (the systems development life cycle approach and the prototyping approach) far outweigh the disadvantages of no system development in the City of New York, Human Resources Administration Department. The automated work are processed with over 100% more efficiency and accuracy, with over 100% more monetary returns and more realization of profits for the City of New York. Better working conditions for users and management was also realized with the introduction of new and better computer hardware and software in the workplace environment where these two systems were developed and implemented and used successfully continuously afterwards, even today and in the future.

The disadvantages that could have been experienced were the continual use of manual processing in these two departments of the City of New York causing very high costs and the incessant possibility of manual errors and inefficient processing of work. This would have also meant long delays and time wasted while waiting for documents to be turned around to the different users for processing.

The advantages of the two systems include the realization of profits and savings and the generation of monetary receipts far above that was received in the 1980s and before for these two departmental units, DCAF and the Office of Revenue and Investigation. This was a trend that was being realized by other parts of the government and also by other businesses and organizations and individuals in the 1990s and beyond.

There are therefore more advantages than disadvantages of developing business computer software and hardware systems using the systems development life cycle and the prototype development method. There are other methods of systems development as described in chapter 2 in the literature review, but these two methods were better suited for the development of the two systems, CAMFR and Bond and Mortgage and the type of users.

## **Discussion**

Conclusions drawn from this case study include the fact that internal MIS computer system workers were available in this company. When these MIS workers are also employees of the company in which the new computer hardware and software systems are being developed there seems to be a change in the outlook of the MIS workers and the users and management. These MIS workers understood what the company was doing and were better able to develop the system than hiring an outside consultant that knows nothing of the company internal job processes. First of all, the consultant would have had to be trained in company procedures and culture in order to begin the systems development project.

Both methods of systems development, the systems development life cycle and the prototyping method work with systems development although there are many other and newer methods of systems development. Typically the systems development life cycle project took more time to develop than the prototyping project.

A significant finding is that these projects were successfully developed and are up and running without the help from MIS staff today and will go on being used successfully in the future. It is expected that there will be some maintenance and enhancements to the systems as the years go by, but the basic systems are strong and form the basis for the development of future redesigned systems.

The theory for systems development is true to the classroom and the workplace also. If the steps in the development methods were followed, then there would be better and more successful systems development. Books, especially systems development books, are not only for school and the classroom they are for the workplace also.

Many parts of the systems development methods were used in this case study, for both the systems development life cycle method and the prototyping development method. Some parts that were heavily included were the design of the two systems, as described in chapter 4, which the MIS staff completed in detail for user approval, and then the systems implementation took place.

The highlights of the CAMFR system were as follows. With the work of the CAMFR system, the work processing was to a major extent redesigned to take advantage of new technologies such as telecommunications technology. The instant transfer of transaction files from a front-end Unix LAN system to a back-end IBM mainframe and the instant printing of reports from the back-end mainframe to the front-end user site were the main highlights of the redesigned CAMFR system. Before, it took days before these two transfers were completed by the use of magnetic sequential tapes and trucks travelling between the two locations.

The highlights of the Bond and Mortgage system were as follows. The use of IBM Pentium computers to process the work and also the use of Microsoft Office 97 Professional software to develop a database, data-entry forms, Word documents, Excel spreadsheets, reports, appointment scheduling, and accurate and extensive storage of records. The user was then able to concentrate on the task of collecting the monies due the City of New York without being handicapped with non-effective hardware and software such as one 286 DOS standalone computer and paper records and rows of filing cabinets with files of records. With the new system, records were retrieved in light seconds by key fields such as Case Number instead of having to search and rummage through draws of misfiled records trying to search for a lien holder and trying to trace the processing of the collection of monies from the clients.

This study showed the importance that no matter what type of system development used to develop and implement a system, in the end it is the procedure followed to ensure proper development and successful implementation and use of the system. The bottom line was that the managers had to decide which method of systems development was most suitable for the development of a particular system. This was evident in the fact that the managers of the CAMFR system had completed their feasibility study and was ready for a detailed redesign of the CAMFR system that they expected to take many years. The managers of the Bond and Mortgage system were about to purchase IBM Pentium computers using Windows software for their unit and therefore wanted a prototype system developed and implemented for their users who were all about to receive Pentium computers to do their daily work. The Bond and Mortgage system managers after doing the feasibility study expected the system to be developed in one to two years.

The managers were pro-technology. The managers believed that information technology is vital to the business and they wanted to use the technologies and new computer hardware and software extensively. This was aided by the fact of the rapid decrease in price for computers and computer software in the 1990s and later. Businesses began experiencing huge profits gained from the cheaper and more efficient processing of work done on less expensive computer hardware and software available from since in the 1990s and in later years.

Since this researcher, the systems analyst in the case study, was the main developer of the system designs for the CAMFR system and Bond and Mortgage system, the documentation was developed and kept by this researcher. Therefore, permission was not needed for the figures in Appendix A of the CAMFR system user manual and figures in Appendix B of the Bond and Mortgage sample merged Word document from records in the Access database.

## Recommendations

This case study included the use of internal personnel for systems development projects instead of hiring consultants from the outside. The trend today is for many companies to hire outside consultants to do their computer systems development projects. This came about ever since the Year 2000 coding problem in the late 1990s. Future research could compare a project developed internally and a project developed externally with external workers. The following aspects could be measured: if there is any difference in the method of development; which system the users understood better; which system cost the least to develop; and the amount of time spent in the development of the two projects.

Other studies could also be done with other systems development approach methods and a measure of the success of the implemented system or failed system could be analyzed. System development methods such as object oriented development and neural network system development are prime methods for future case studies. Also, newer inventions of technology could be incorporated into new projects in order to develop more enhanced 21st century systems that could last for centuries.

The implication of this research to the discipline of Management Information Systems is that good projects can be developed successfully and could work successfully for many years into the future if the theory for systems development methods is followed. This is so because these systems development methods have been tried and proven.

This study will contribute to our knowledge about computer systems development in that it shows that computer hardware and software systems could be developed to replace the manual work done by employees. This would produce more efficient projects and job processing and would produce more profits for the company. This in turn would benefit the nation.

### Conclusion

Any method of system development used for the computerization or automation of work processes is acceptable as long as it produces a working computer system that fit the users and management needs. Developed computer systems improve the way work is done so much so that efficiency, profits, and production of more and superior products are realized. This is illustrated in this study by the development and implementation of the two computer software systems, CAMFR and Bond and Mortgage, which were developed using the system development life cycle method and the prototyping method respectively. The City of

New York would not have experienced profits and efficient work production in these two departments if the two systems were not developed to take the place of old inefficient systems and manual systems. In this day and age of technology advancement and inexpensive computer software and hardware in relation to their capabilities, every business, government, and organization should develop business computer systems to carry out daily operations better.

# BIBLIOGRAPHY

Gossain, Sanjiv. <u>Object Modeling and Design Strategies</u>. Cambridge University Press, 1998.

Larman, Craig. <u>Applying UML and Patterns. An Introduction to Object-Oriented Analysis and Design</u>. Prentice-Hall, 1998.

Leedy, Paul D. and Jeanne Ellis Ormrod. <u>Practical Research. Planning and Design</u>. Seventh Edition. Prentice-Hall, 1985.

McClave, James T., P. George Benson, and Terry Sincich. <u>Statistics for Business and Economics</u>. Seventh Edition. Prentice-Hall, 1998.

McLeod, Jr., Raymond. <u>Management Information Systems</u>. Sixth Edition. Prentice Hall, 1995.

McNurlin, Barbara C. and Ralph H. Sprague, Jr. <u>Information Systems Management in Practice</u>. Fourth Edition. Prentice-Hall, 1998.

Meriwether, Nell W. <u>Successful Research Papers in 12 Easy Steps</u>. NTC Publishing Group, 1997.

Peters, Thomas J. <u>Thriving on Chaos, Handbook for a Management Revolution</u>. Alfred A. Knopf, Inc., 1987.

Post, Gerald V. and David L. Anderson. <u>Management Information Systems. Solving Business Problems with Information Technology</u>. Second Edition. McGraw-Hill, 2000.

Sprague, Jr., Ralph H. and Hugh J. Watson. <u>Decision Support for Management</u>. Prentice Hall, 1996.

Turban, Efraim and Jay E. Aronson. <u>Decision Support Systems and Intelligent Systems</u>. Fifth Edition. Prentice-Hall, 1998.

# APPENDIX A

# CAMFR SYSTEM SCREENS

## (Excerpts from the User Manual)

*Dr. Jennifer W.  Gilmore, PhD*

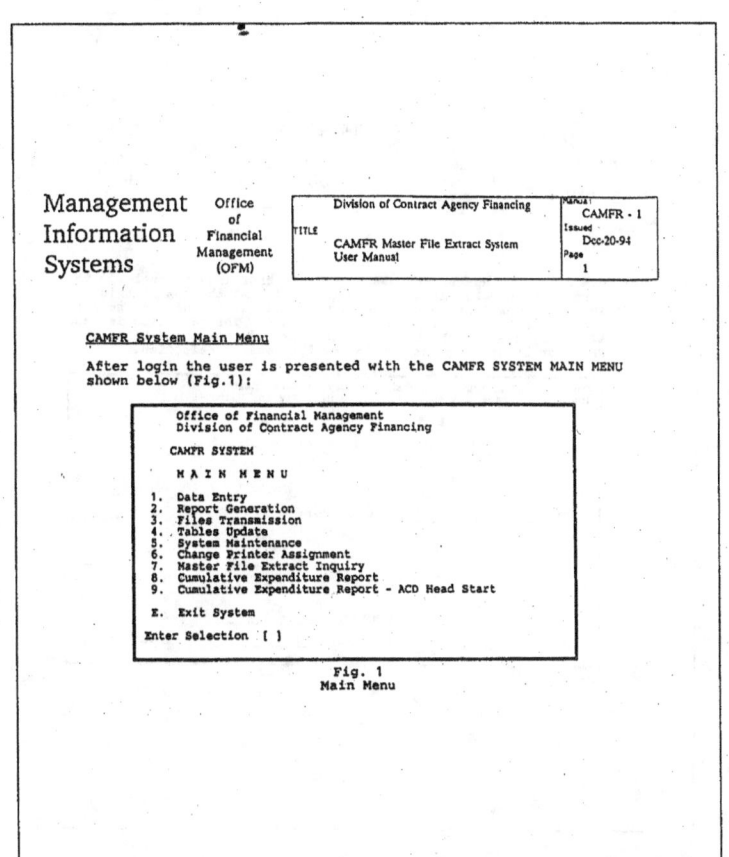

Figure 1: The Main Menu of the CAMFR System user data-entry screen.

| Management Information Systems | Office of Financial Management (OFM) | Division of Contract Agency Financing | | CAMFR - 1 |
|---|---|---|---|---|
| | | TITLE CAMFR Master File Extract System User Manual | | Issued Apr-6-93 |
| | | | | Page 2 |

Pressing 7 brings up the Master File Extract Inquiry System. The screen shown below is displayed (Fig. 2). At the top of this screen is the Primary Ring Menu, which allows the user to select the Contract, Financial or History screen. The selection is made either by placing the cursor on the desired choice and hitting ENTER or by hitting the Initial of the desired selection.

```
Options: Contract, Financial, History, Report, Exit
Inquire Contract Extract File (Press Help Key for Help)

                    Division of Contract Agency Finance
                            Master File Extract
                       Contract Summary Screen (as of___ )

                                                                JES:

  ID: _____        START: __/__  END: __/__   HOLD: _____

  NAME: _____        LAST CAMFR: __/__          BUD: _____

  STR: _____        CAMFRS RCVD: ___

  CITY: _____        STATE: __ ZIP: _____       OBJ: _____

                            CONTR AMT: _____        SUB: _____

  REG #:
  IRS #:                    CUM EXP: _____          ORIG: _____

  BUD LINES: _____          CLOS BAL: _____         CL CAT: _____

                            Fig. 2
```

Figure 2: CAMFR System Master File Extract Contract Summary Screen.

| Management Information Systems | Office of Financial Management (OFM) | | | |
|---|---|---|---|---|
| | | Division of Contract Agency Financing | | Manual CAMFR - 1 |
| | | TITLE CAMFR Master File Extract System User Manual | | Issued Apr-6-93 |
| | | | | Page 3 |

Contract Summary Screen

Hitting C (Contract) on the Ring Menu will display the screen below (Fig. 3):

```
                                                      Ctrl-X to Abort
Enter Search Criteria and Press <ESC> Key
contring.inq         Division of Contract Agency Finance
                            Master File Extract
                     Contract Summary Screen (as of___ )

                                                           JE5:

  ID: _____   START: __/__  END: __/__   HOLD: ____

  NAME: _____  LAST CAMFR: __/__          BUD: ____

  STR: _____   CAMFRS RCVD: ___

  CITY: _____   STATE: __ ZIP: _____    OBJ: ____

                        CONTR AMT: _____       SUB: ____

  REG #:
  IRS #:               CUM EXP: _____         ORIG: ____

  BUD LINES: ____      CLOS BAL: _____        CL CAT: ____
```

Fig. 3
Contract Summary Screen

The user can enter search information in any of the fields on this screen (except City, State and Claim Category). The system will select all records that meet the search criteria. The user can move the cursor from field to field by pressing the ENTER key. After all entries are made, hit *Ctrl* and [ simultaneously. This will be referred to as Ctrl + [ in the following sections.

Sample Use: Move the cursor to the ID field, key in a valid contract number and Hit Ctrl + [ to obtain all records with this number. There will be only one record returned with this number.

Figure 3: CAMFR System Master File Extract Contract Summary Screen with user instructions.

131

```
┌─────────────────────────────────────────────────────────────────────┐
│                            ♣                                          │
│                                                                       │
│ Management      Office    ┌──────────────────────────────┬──────────┐│
│                   of      │  Division of Contract Agency  │ CAMFR - 1││
│ Information    Financial  │          Financing            │ Issued   ││
│                Management │ TITLE                         │ Oct-12-94││
│ Systems          (OFM)    │  CAMFR Master File Extract    │ Page     ││
│                           │  System User Manual           │   4      ││
│                           └──────────────────────────────┴──────────┘│
│                                                                       │
│   The system returns the Contract Summary screen as shown below       │
│   (Fig. 4):                                                           │
│                                                                       │
│   ┌─────────────────────────────────────────────────────────────┐   │
│   │ Options: Contract, Financial, History, Report, Exit          │   │
│   │ Inquire Contract Extract File (Press Help Key for Help)       │   │
│   │                                                               │   │
│   │ contring.inq       Division of Contract Agency Finance        │   │
│   │                         Master File Extract                   │   │
│   │                 Contract Summary Screen (as of 12/7/1992)     │   │
│   │                                                        JE5:   │   │
│   │ ID: 6241190120001016    START: 07/92  END: 06/93  HOLD: *016  │   │
│   │ NAME: # 016 Brevoort CC  LAST CAMFR: 10/92        BUD:  3703  │   │
│   │ STR: 1185 Park Place    CAMFRS RCVD: 4                        │   │
│   │ CITY: Brooklyn          STATE: NY ZIP: 11233      OBJ:  552   │   │
│   │                         CONTR AMT: 318911         SUB:   1    │   │
│   │ REG #:                                                        │   │
│   │ IRS #:                  CUM EXP: 116288      ORIG: 069        │   │
│   │ BUD LINES: 11           CLOS BAL: 39342           CL CAT: 3016│   │
│   └─────────────────────────────────────────────────────────────┘   │
│                            Fig. 4                                      │
│                                                                       │
│   As another example, if 11233 is entered in the ZIP field, all       │
│   contracts with this zip code will be selected.  A browse menu       │
│   will appear on the screen (see Fig. 5).  This menu allows the       │
│   user to navigate through the selected contracts in one of several   │
│   ways.  Browse menu definitions are shown below.                     │
│                                                                       │
│   Next          Displays the next record in sequence                  │
│   Previous      Displays the previous record                          │
│   First         Displays the very first record selected               │
│   Last          Displays the very last record selected                │
│   Scrn Prt      Prints whatever is on the screen at the time          │
│   Exit Browse   Brings back the Primary Ring Menu.                     │
│                                                                       │
└─────────────────────────────────────────────────────────────────────┘
```

Figure 4: CAMFR System Master File Extract Contract Summary Screen filled in and with user instructions.

```
                                        ≗

   Management      Office        ┌──────────────────────────────┬──────────┐
                      of         │  Division of Contract Agency Financing │ Manual   │
   Information     Financial     │                                │ CAMFR - 1│
                   Management     │ TITLE                          │ Issued   │
   Systems          (OFM)        │     CAMFR Master File Extract System │ Oct-12-94│
                                 │     User Manual                │ Page     │
                                 │                                │   5      │
                                 └──────────────────────────────┴──────────┘


   ┌──────────────────────────────────────────────────────────────┐
   │ BROWSE:  Next   Previous   Last   First   Scrn-Prnt   Exit-Browse │
   │ Type a Number (the skip value) Before Entering N To Skip That Many Rows │
   │ ------------------------------------------------------------- │
   │ contring.inq        Division of Contract Agency Finance        │
   │                          Master File Extract                   │
   │                  Contract Summary Screen (as of 12/7/1992)     │
   │                                                         JE5:    │
   │ ID:  6241190120001016     START: 07/92  END:  06/93  HOLD: *016 │
   │ NAME: # 016 Brevoort CC   LAST CAMFR: 10/92          BUD:  3703 │
   │ STR:  1185 Park Place     CAMFRS RCVD: 4                        │
   │ CITY: Brooklyn            STATE: NY ZIP: 11233       OBJ:  552  │
   │                           CONTR AMT:   318911        SUB:   1   │
   │   REG #:                                                        │
   │   IRS #:                  CUM EXP:   116288          ORIG: 069  │
   │   BUD LINES: 11           CLOS BAL:  39342           CL CAT: 3016│
   │ Record 20 of 28                Total of Contr Amt = $10,152,310 │
   └──────────────────────────────────────────────────────────────┘
                              Fig. 5
                  Browse Mode of Contract Summary
```

Figure 5: CAMFR System Master File Extract Contract Summary Screen in Browse Mode.

| Management Information Systems | Office of Financial Management (OFM) | Division of Contract Agency Financing | CAMFR - 1 |
|---|---|---|---|
| | | CAMFR Master File Extract System User Manual | Issued Apr-6-93 |
| | | | Page 6 |

Financial Screen

Hitting F in the Ring Menu will bring up the screen shown below (Fig. 6).

```
finaninq.per              Master File Extract
                    Financial Summary Screen (As of ___ )

ID                    CONT AMT      START      END      BUD LINES
                                      /         /

BUD CD       BUD AMT      CUM EXP      UNPD BILLS      PROJECTION



TOTAL - 1100:
TOTAL - 1200:
Enter all 16 digits of Contract ID.  Ctrl-Z to Abort.
```

Fig. 6
Financial Summary Screen

Figure 6: CAMFR System Financial Summary Screen.

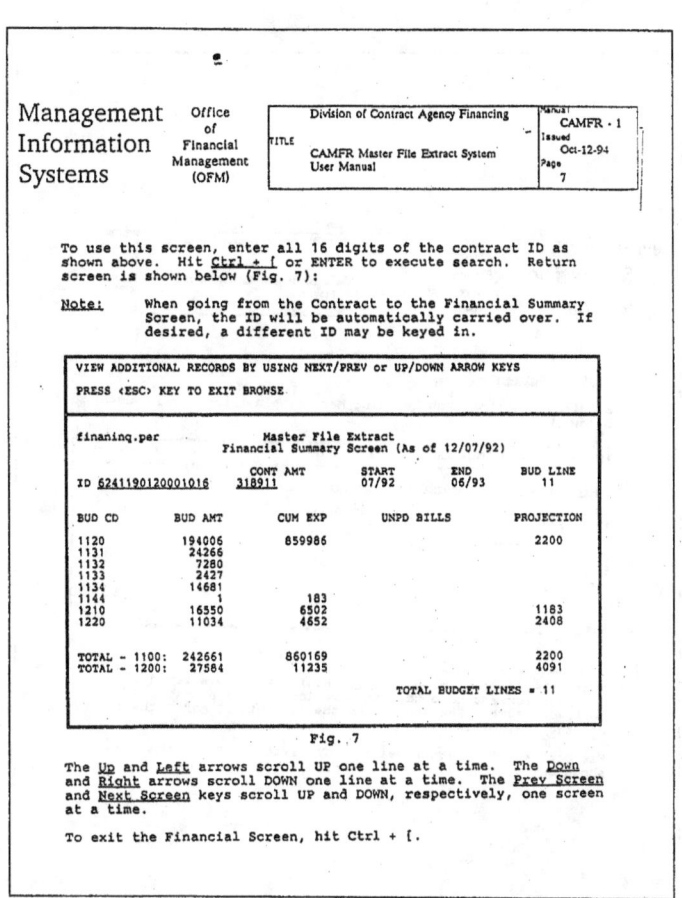

Figure 7: CAMFR System Financial Summary Screen filled in with user instructions.

| Management Information Systems | Office of Financial Management (OFM) | Division of Contract Agency Financing | | Manual CAMFR - 1 |
|---|---|---|---|---|
| | | TITLE | CAMFR Master File Extract System User Manual | Issued Oct-12-94 |
| | | | | Page 8 |

### History Screen

Hitting H (History) on the Ring Menu displays the screen shown below (Fig. 8):

```
ENTER SEARCH CRITERIA AND PRESS <ESC> KEY              Ctrl-X to Abort
------------------------------------------------------------------------
histinq.per      DIVISION OF CONTRACT AGENCY FINANCE
                      MASTER FILE EXTRACT
                        HISTORY SCREEN
      ROOT ID
ID _ _____  _          AGCY

                             IRS

START  /            END    /        AMT
```

Fig. 8
History Screen

Enter the 14 digit root ID, not the 16 digit ID entered on previous screens to obtain information for the previous 2 contract years.  The Root ID is the 16 digit ID minus the first and the last digits.

Hit Ctrl + [ to execute search.  The system returns the screen shown below (Fig. 9):

Figure 8: CAMFR System History Screen with user instructions.

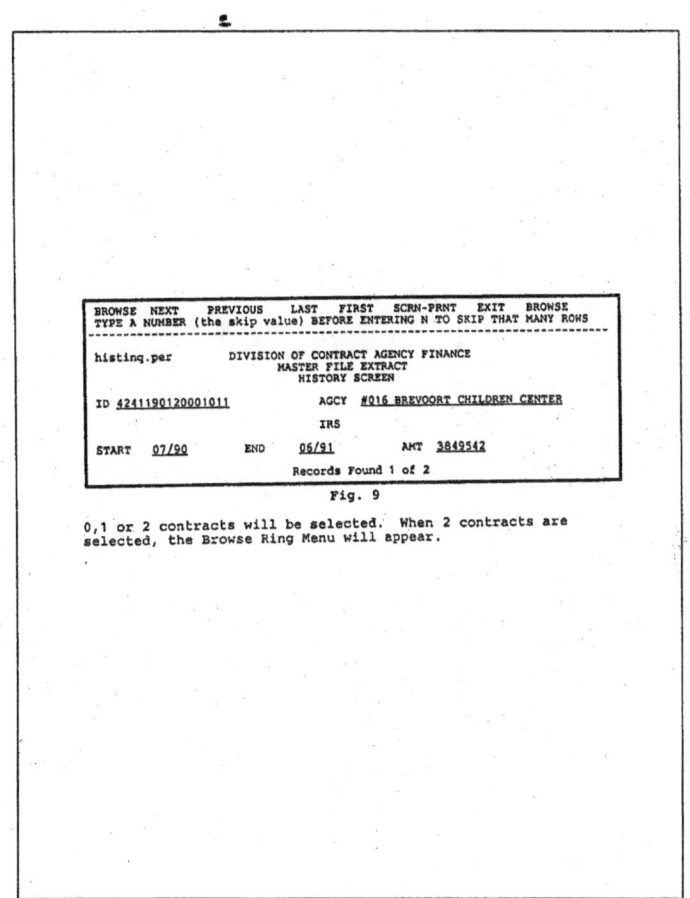

Figure 9: CAMFR System History Screen filled in with user instructions.

# APPENDIX B

## A BOND AND MORTGAGE SYSTEM MERGED MICROSOFT WORD DOCUMENT WITH DATA FROM THE MICROSOFT ACCESS DATABASE

HUMAN RESOURCES ADMINISTRATION
OFFICE OF REVENUE AND INVESTIGATION
60 HUDSON STREET, 8TH FLOOR
NEW YORK, NY 10013

GABRIEL W. GORENSTEIN                    ALBERT M. GIOVE
*General Counsel*                         *Deputy Administrator*

                                  Date: ____10/22/97___

TO:      ____«UNITSEC»_____
         Unit/Section
         ____«WORKERNAME»_____
         Worker

FROM:    Real Property and Asset Unit

         __«RPWORKER»_____  «RPPHONE»____
         Worker             Telephone #

                         Re:_«LASTNAME»_«FIRSTNAME»_
                             Case Name
                             ____«CASE»_____
                             Case Number

This file is being returned for the reason(s) indicated:

[    ]   No response to request for information.

[    ]   Reason for referral missing on the MAP-2109 Referral.

[    ]   This property shows a net deficit – no rental income to be considered (W-506
         Prepared).

[    ]   This property shows a net annual income of $_«NETPINCOME»_.
         (W-506 Prepared).

[    ]   Other        _____
         (Specify)

Comments:

_____
_____
_____

Figure 10: A Bond and Mortgage System Microsoft Word Document with merge
fields.

HUMAN RESOURCES ADMINISTRATION
OFFICE OF REVENUE AND INVESTIGATION
60 HUDSON STREET, 8TH FLOOR
NEW YORK, NY 10013

GABRIEL W. GORENSTEIN                    ALBERT M. GIOVE
*General Counsel*                        *Deputy Administrator*

                              Date:    ___10/22/97___

TO:    ___FH___
       Unit/Section
       ___T. WALLACE___
       Worker

FROM: Real Property and Asset Unit

       __L. KIMBLE_____2122745600___
          Worker        Telephone #

                              Re:_MERVIN, CHRISTIE___
                                   Case Name

                              26186705
                              Case Number

This file is being returned for the reason(s) indicated:

[    ]   No response to request for information.

[    ]   Reason for referral missing on the MAP-2109 Referral.

[    ]   This property shows a net deficit – no rental income to be considered (W-506
         Prepared).

[    ]   This property shows a net annual income of $_$0.00_.
         (W-506 Prepared).

[    ]   Other    _____
         (Specify)

Comments: _____

_____

**Figure 11**: A Bond and Mortgage System Microsoft Word Document with merge fields filled in with data from the Microsoft Access Database.

# ABOUT THE AUTHOR

Dr. Jennifer Williams Gilmore, Ph.D. is a Project Manager and Computer Specialist at the City of New York, Human Resources Administration, Management Information Systems Department. She is also an Adjunct Computer Information Systems Professor currently teaching at Monroe College and Touro College in New York City, and an Online Adjunct Full Professor at the University of Maryland University College. Dr. Gilmore holds two Ph.D. degrees from Walden University and Kennedy-Western University, both were received in 2001. At Walden University her major was Applied Management and Decision Sciences with a self-designed specialization in Management Information Systems. At Kennedy-Western University her major was Management Information Systems. Dr. Gilmore also holds many other degrees as follows: an MBA in Management from Long Island University (Brooklyn) (1994) and an MS in Computer Methodology from Baruch College (CUNY) (1993). She also holds a BBA in Computer Information Systems from Baruch College (CUNY) (1986); an MA in Economics from Brooklyn College (CUNY) (1984); a BA in Economics from Brooklyn College (CUNY) (1978); and an AS in Liberal Arts and Sciences from New York City College of Technology (CUNY) (1974). Dr. Gilmore has taught as an Adjunct Lecturer in many colleges, as follows: New York City College of Technology, Kingsborough Community College and the Borough of Manhattan Community College. She also taught at Medgar Evers College, St. Francis College (Brooklyn), and Baruch College. Dr. Gilmore migrated from Trinidad, West Indies in 1972 and became a US citizen in 1993. She currently lives in Brooklyn, New York.